Communicating Science

an introductory guide for conveying scientific information to academic and public audiences

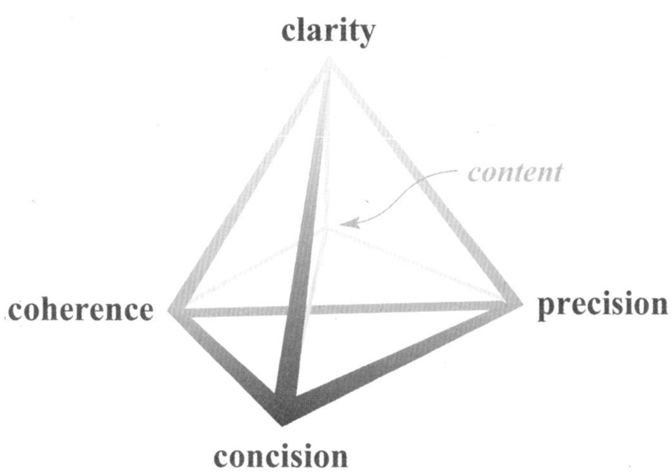

Roy Jensen

Communicating Science: an introductory guide for conveying scientific information to academic and public audiences

ISBN 978-0-9782390-8-4

Print edition. This resource provides you with a solid foundation in science communication. Use this resource in all your courses and as a reference guide in your career.

If you have any comments or questions, please use the Contact link on RoguePublishing.ca

Versions of this resource

ISBN 978-0-9782390-8-4 (print edition)
ISBN 978-0-9782390-9-1 (learner electronic edition)
ISBN 978-0-9782390-7-7 (instructor electronic edition)

Library and Archives Canada Cataloguing in Publication

Jensen, Roy, 1971-, author
Communicating Science: an introductory guide for conveying scientific information to academic and public audiences / Roy Jensen.
Includes index.
Issued in print and electronic formats.
ISBN 978-0-9782390-8-4 (pbk.) – ISBN 978-0-9782390-9-1 (pdf) –
ISBN 978-0-9782390-7-7 (pdf)
1. Communication in science. I. Title.
Q223.J45 2014 501'.4 C2013-908234-4
C2013-908235-2

Printed in North America.
10 9 8 7 6 5 4 3 2

Contents

Preface

Communicating Science provides undergraduate science and engineering students and new technical writers with a foundation for writing, reviewing, and presenting scientific information: reports, proposals, scholarly articles, essays, theses, scholarly posters, oral presentations, and documents for public audiences.

Somewhat surprisingly, writing guidelines are not uniform throughout the sciences and even vary within a scientific discipline. While many scientific organizations publish professional style guides, these guides are high-level documents written for scientists, authors, editors, and publishers. These guides can overwhelm a new writer. *Communicating Science* consolidates common communication concepts for new science writers. *Communicating Science* is unique in its breadth, exploring a wide array of documents that a science writer may be required to prepare and present and contains extensive sections on peer review and editing. *Communicating Science* provides graduate students and/or professional writers with a solid foundation for using the professional style guides of their chosen career.

Communicating Science explores all aspects of preparing quality professional documents and presentations.

Chapter 1 (Fundamentals of communication) presents strategies for effective communication of scientific information and reviews grammar and style expectations of academic documents and presentations.

Chapter 2 (Communicating scientific information) focuses on the formatting of scientific information, tables, and figures, and expectations when citing information.

Chapter 3 (Fundamentals of learning) introduces the constructivist learning model and explores the biophysical processes that occur during learning. Knowing how people learn is important when preparing quality documents and presentations.

Chapter 4 (Research methodology) discusses the process of conducting research and investigative projects and introduces research methods and research ethics.

Chapter 5 (Documents and presentations) presents guidelines for producing and formatting documents and presentations commonly required of scientists and academic science writers.

Chapter 6 (Peer review and peer evaluation) presents tools and strategies for engaging in effective peer review and peer evaluation, including strategies for giving and receiving feedback. This chapter emphasizes the importance of such skills in preparing quality documents and presentations.

Appendix A (Review questions) lists questions authors should keep in mind to guide the development of documents and presentations. Reviewers should use these questions when reviewing documents and presentations.

Appendix B (Electronic document preparation) instructs authors on how to use the Microsoft Office® suite of tools to produce professional-quality documents and presentations.

Appendix C (Assignments) presents assignments that build skills, experience, and confidence in preparing and presenting documents. Most assignments can be augmented with peer review and/or peer evaluation.

Appendix D (Assessment rubrics) presents rubrics for evaluating the content, organization, and presentation (if applicable) of laboratory reports, scholarly articles and essays, scholarly posters, and scientific presentations. A rubric for evaluating the reviewer during peer review activities is also included.

While *Communicating Science* does progress from simple to complex concepts, it is possible (and valuable) to teach the chapters in a different order. For example, applying the concepts in Chapters 1 and 2 to the documents in Chapter 5 results in students concurrently improving their grammar, presentation of scientific information, and their ability to compose scientific documents and presentations. Additionally, teaching the editing and peer review components in Chapter 6 early makes peer review integral to the preparation of documents and presentations.

Students can use *Communicating Science* throughout their undergraduate program. Depending on the institution and program offerings, there are several ways of integrating *Communicating Science* into the undergraduate program:

- as the course textbook for a science or technical communication course
- as a resource that is used in many courses from first- to fourth-year
- as a resource for students engaged in undergraduate research

Communicating Science is also a valuable resource for persons entering careers where they are required to prepare quality professional documents and presentations.

Using this resource

To get the most out of this resource, you must apply the techniques herein in your everyday activities. Remain conscious of grammar rules and tenses, and think about how to improve the readability of your work. This will improve your writing style, whether you are composing an email, résumé, or scientific document. Get into the habit of reviewing your work and find colleagues with whom you can review each other's work. This ensures your final documents and presentations are of the highest quality.

The **Additional resources …** sections list the resources consulted while preparing each chapter and additional resources that may be relevant for anyone wanting to learn more about a given concept. In some sections, information was adapted directly from other sources, and these resources are cited appropriately.

Acknowledgements

I must express my thanks to the following people who have reviewed this document so that it will be of the most value to you:

Aaron Swanbergson
Audrey Habke
Belinda Ongaro
Caitlin Guzzo
Caroline Lee
Dr. Craig MacKinnon
Erin Dul
Dr. Hugh Cartwright
Dr. Julie B. Ealy
Dr. Lawrence McGahey
Niall Fink
Dr. Umesh Parshotam

I am especially indebted to Sarah-Nelle Jackson for her assistance with Chapter 1 and with editing of the entire document.

A final thanks goes to Dr. Jeremy Tatum. Dr. Tatum sparked my interest in science communication as he was identifying grammatical and style errors in my dissertation … it is amazing where interests begin.

Comments and suggestions to improve this document are appreciated!

Roy Jensen

Offer me a choice between two entrants, [both curious about science, motivated, and ambitious.] One has inadequate chemistry but good literacy skills, the other the converse. Which do I feel has the greater chance of succeeding in a science degree and beyond? Well, I shall take the literate one. Those who disagree with me are quite welcome to the other. — Peter Towns

Chapter 1. Fundamentals of communication

Communication is a critical part of existence. We communicate with our families, friends, co-workers, clients, and with strangers. *Verbal* and *non-verbal* communication are the dominant forms of communication with those around us. *Written* communication is the dominant form with those distant from us and used to create lasting records. Meetings and discussions are often followed-up with a written document detailing the topics and decisions. Being able to prepare these documents is critical to success in your education, career, and personal life, and is the focus of *Communicating Science.*

In science, the goal is to unambiguously convey accurate and detailed technical information to others, including others whose first language is not English. Spelling errors, grammatical errors, and the phrasing of your prose may cause readers to misinterpret the information you intend to convey, question the quality of the science you conducted, and question your abilities in general. When applying for employment, the decision to interview you is based primarily on your written application.

Improving your writing skills also improves your thinking skills by forcing you to think logically and critically, which makes you a better and more effective communicator. Furthermore, communication is bidirectional: being an effective communicator increases your ability to comprehend what is communicated to you.

Figure 1.1 The components of effective communication.

1.1 Elements of effective communication

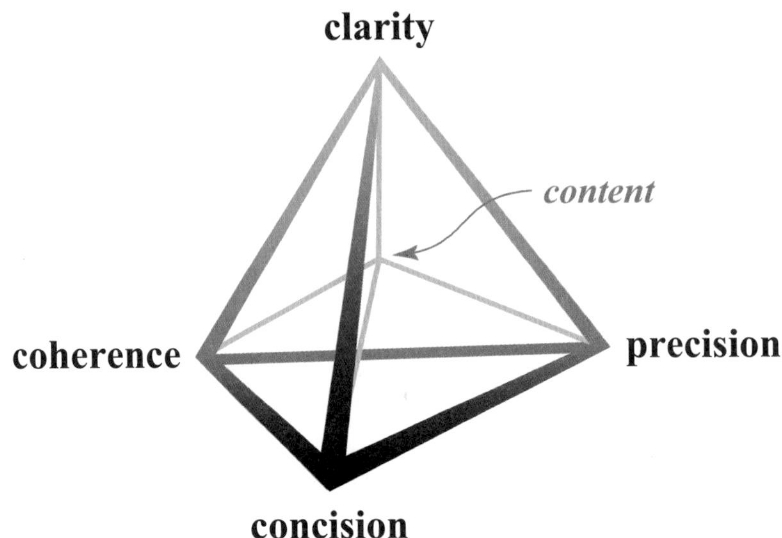

Figure 1.2 Hallmarks of quality scientific prose that facilitate the communication of content (information).

A critical component of any communication is that there must be *content* (*information*) to communicate. Without content, you are saying nothing and the reader and/or audience quickly realizes that. Once you have information to communicate, convey it with clarity, coherence, concision, and precision so the reader obtains the greatest possible knowledge from the information you share.

Clarity refers to clear and unambiguous prose that is suited to the reading level and the background of the reader. The reader should obtain the desired understanding without having to reread sections of the document.

Coherence refers to organization and structure that presents information in a logical order.

Concision refers to brief but complete prose. The writer conveys the information in as few words as possible, without repetition. Include information that relates to the topic and exclude extraneous information.

Precision* refers to the selection and use of the correct words to convey the information and obtain the desired understanding.

* Another definition of *precision* pertains to uncertainty in scientific measurement. This second definition is used later in *Communicating Science* when discussing data analysis.

Information becomes *knowledge* when it is understood by others.

Literary and academic works*

Literary works include novels, short stories, poetry, plays, and movies. Literary works are valued for their subtlety, multiple interpretations, and ability to excite the reader's imagination.

Academic works include non-fiction books, encyclopedia articles, news articles, reports, theses, scholarly articles, presentations, and documentaries. Academic works are valued for their accuracy, focused singular message, and ability to interest and engage the reader.

You practice and hone the skills of composing literary and academic works by preparing these works and by reviewing the works of others. However, students and scientists have, historically, had limited instruction in preparing academic works.

Like literary works, academic works must interest the reader so that they want to know the complete story — not the fictional story of a literary work — but the factual story of your research/investigative project. Your academic work must

- be truthful (no false or fabricated information)
- be focused on your topic (no extraneous information)
- flow smoothly and logically from idea to idea
- interest and engage the reader

Scientific academic works commonly answer four questions:

1. Why was research conducted? (background, introduction)
2. What research was conducted? (research question, method)
3. What knowledge was gained? (results, discussion, conclusion)
4. What more can be done? (future work)

Answering Question 1 explores the limitations or inconsistencies in the existing knowledge, which sets the stage for a research/investigative project (Question 2). Answering Question 3 explains what was done and how it improves and expands our understanding in this scientific field. Since advancements in science are incremental, Question 4 proposes a course of action to continue the work.

* *Work* is a general term that includes written documents and presentations. See section 2.3 for details.

Writing for the reader

When writing, always consider the readers:

- What information are readers interested in?
- What foundational knowledge do readers have?
- How can you keep readers interested and engaged?

Once you have these answers, you can produce a coherent document that conveys the desired information to the reader in a clear and concise manner. Common strategies to interest and engage the reader include

- writing at the level of the reader
- writing as if you are talking to the reader
- writing in a positive and professional tone
- writing to persuade the reader of your ideas and conclusions

A common error of science writers is assuming that readers have more knowledge than they actually do. Scientists are often surprised to learn how little scientific knowledge the public actually has, and struggle to convey information using terms understood by the public. You may have experienced this if you have tutored someone in math or science: you were surprised when the tutee did not know something that "everyone knew" and were even more surprised when the tutee struggled to learn that concept.

Tutor: "Don't you know the order of mathematical operations?"

Oral communication is approximately 10 % verbal (the words you are saying), 60 – 80 % non-verbal (the tone and emphasis of your words, your posture, facial expressions, gestures, etc.), and 10 – 30 % your appearance (your hair, clothing, cleanliness, etc.). Written communication loses 90 % of the message, but you must still convey the information and your interest and enthusiasm for the project in the words you choose. The written words must say exactly what you want to communicate. This is especially important because you may be communicating scientific information to readers who do not have the same scientific or cultural background or whose first language is not English.

Since scientific information is complex, you want the prose to be as simple and clear as possible. The following strategies improve the readability of your writing for all readers.

- *Clarity*: use the correct words; use the primary meaning of words; use words appropriate for the audience; use simple tenses; make prose understandable on the first reading; use figures, images, and graphs to facilitate understanding.
- *Coherence*: build a story for the reader; present information when required and in a logical progression; do not make the reader jump to other sections of the document.
- *Concision*: minimize the number of words; ensure every word, sentence, and paragraph adds meaningful information; use simple phrases and sentences; use the active voice wherever possible.
- *Precision*: say exactly what you mean; use correct grammar and style; be numerically exact when possible.
- Define technical terms and abbreviations, and minimize the use of scientific jargon (technical terminology understood only by experts).

> *Vagueness, ambiguity, and inability to express clearly and succinctly are intolerable in a scientist.* — Peter Towns

Tone in communication

Tone is the attitude and emotion that you convey in the words you use and the emphasis you apply to those words. When speaking, listeners have both the words and your non-verbal oral cues (pauses, vocal inflections, emphasis) to identify the tone you intend for the information. When reading, readers only have your words and must infer tone from those words.

> *It is not what you say; it is how you say it.* — unknown

In literary works, writers set any number of tones: angry, authoritative, condescending, courteous, enthusiastic, exciting, hurt, inviting, playful, patronizing, romantic, sad, sincere, threatening, etc.

Consider the following sentences:

The kites danced in the wind and sun.	(joyful, happy)
The wind and waves raged against the shore.	(scary, dangerous)
The light breeze drifted across the quiet beach.	(tranquil, lonely)
The old house moaned and creaked in the cold wind.	(spooky, isolated)

Depending on whom you are interacting with and why, your tone could be casual (with friends) or formal (with professionals).

How's it going?	(casual)
How are you today?	(formal)
Why hasn't the prof returned our midterms?	(casual, demanding)
I wonder why the professor has not returned our exams.	(formal, inquiring)
So, what's going on here?	(casual, inquiring)
What the hell are you doing!?	(angry, aggressive)

Readers are exceptionally good at picking up tone from prose. Consider the range of emotions you experience reading well-written novels. Readers expect academic works to convey positive, formal, objective, and confident tones. Together, these qualities embody *professionalism* in the work. If the tone is not what readers like or expect, they will not want to read the work.

Academic writers focus on their work. They dedicate years to investigating and understanding phenomena. When they write, academic writers often focus on the information they want to convey and do not consider the readers. This introduces a range of random tones: authoritarian, condescending, indifferent, and/or patronizing tones are common, and are exactly what the reader does not want to hear and feel. It is critical for academic writers to always consider the reader when writing and to impart positive, formal, objective, and confident tones in their prose.

It is difficult to judge tone when writing. However, by *reading your work aloud*, you will notice the attitude and emotion being conveyed. As you read your work, consider the following:

- What emotions am I feeling as I read and hear this?
- Would I like this information presented to me in this way?
- How could I convey this information better?

Communicating Science teaches you to create professional documents and presentations. The examples below should help you establish positive, formal, objective, and confident tones in your prose.

Positive tone: express information, even negative information, in a positive light; be polite and courteous.

Your results are wrong because you did not calculate the ____ correctly.	(negative)
The results are incorrect because of an error calculating the ____.	(positive) ✓

Formal tone: use proper capitalization and punctuation; write in complete sentences; use words appropriate to the readers; sound natural, as if you are talking with a professional colleague.

You'll need to replace the batteries or it'll die.	(casual)
Replace the batteries so the meter will work for the duration of the experiment.	(formal) ✓
Replace the batteries.	(demanding)

Objective tone: avoid bias; be honest; do not be swayed by personal or public beliefs; base conclusions and recommendations on facts.*

We do not believe the results because they are not consistent with our expectations.	(biased)
We have reviewed our method and calculations, and now believe that our hypothesis is incorrect.	(objective) ✓

Confident tone: believe in your research and results; believe in yourself; use clear and concise language to convey information.

The broken hinge on the instrument may have affected our results.	(weak)
Our results were affected by a broken hinge on the instrument.	(confident) ✓
You will agree that I am the best candidate for this position.	(arrogant)
Given my education and experience, I am confident I will excel in this position.	(confident) ✓

Strategies for creating documents with positive, formal, objective, and confident tones include:

- Write as if you are talking to the reader.
- Use pronouns (I, me, we, us, you, they, them) to connect yourself and the reader to your work, but do not overuse them.
- Read your work aloud as you review it.
- Review your work from the perspective of the reader.
- Envision other ways your prose could be interpreted.
- Have a colleague(s) review it.

When you identify areas with problems, revise your document to negate alternative interpretations and to focus on your intended meaning.

Finally, having the appropriate tone is critical when conveying negative information or preparing correspondence on sensitive or controversial topics.

> *People may not remember exactly what you did or what you said, but they will always remember how you made them feel.* — Maya Angelou

* Objectivity is a fundamental component of all scientific research.

Quality scientific writing

In 1992, Donald Hayes conducted a quantitative analysis of scientific articles compared with newspaper articles.* He discovered that scientific articles contain longer and more complex words and that their complexity increases over time. Hayes proposed this was because science is becoming increasingly detailed and specialized, and the terminology equally so. However, Hayes also found that the sentence and document structure was far from optimal. Scientific articles contained long, complex sentences that were confusing, unclear, and grammatically incorrect.

Why? Often because scientists have little formal training in academic writing. Most scientists develop their writing style from the scientific articles they read and review — which themselves are not well written — thereby perpetuating the cycle of poor writing. Furthermore, some scientists believe they impress others with technical language and complex sentences.

From a more cynical perspective, poor writing can be used to inflate incremental results. Lengthy, impenetrable sentences filled with technical terms, elaborate words, and Latin phrases can be used to suggest that results and research are momentous. The psychological game is that readers who do not understand the prose might assume they are not as intelligent as the scientists who conducted the research and wrote the article. Unfortunately, incomprehensible prose may be ignored by other scientists, hindering the advancement of science. (When reviewing work by others, diligently scrutinize any section of text that is difficult to read.)

Figure 1.3 *Not* the philosophy to have when preparing academic works.

* Hayes, DP. The Growing Inaccessibility of Science. Nature. 1992;356:739–740.

Incomprehensible and boring scientific prose has another significant consequence: it perpetuates the perception that science is only for the intellectuals and is unimportant in society. On the contrary, science is increasingly integrated into society; an understanding of science is critical to understanding the world around us.

> You can often find quality prose in scientific documents written for public audiences, such as *Discover*, *Nova*, and *Scientific American*.

When the subject is difficult, simplicity is the only way to treat it.
— Thomas Jefferson

Science writers must understand and accept that the domain of scientific knowledge is vast. There are an estimated 20 000 active science journals, each publishing hundreds to thousands of articles per year. Every scholarly article adds an important, but minuscule, amount of knowledge to the domain of scientific knowledge, as illustrated in Figure 1.4.

Figure 1.4 The contribution of every published scholarly article to the domain of scientific knowledge.

Strategies for quality writing

Dr. George Gopen and Dr. Judith Swan* investigated how people read and found that readers expect new information to be presented in a manner that establishes a context for the information. In this way, the development of

sentences → paragraphs → sections → document

produces a coherent document. A good writer puts information where the reader expects to find it. The following general strategies will improve your prose.†

Keep the subject and verb close together‡

Sentence length increases when communicating complex information. There is a tendency to put all the supporting information close to the subject, thereby separating the subject and verb. Do not do this! Keep the subject and verb close together. By telling the reader what the subject did, the reader has the context for why the supporting information is important.

> *Original*: Scientists at CERN, the world's largest particle physics laboratory, funded collaboratively by twenty European countries, have announced the detection of the Higgs Boson, which is the final particle proposed by the Standard Model of particle physics.

The subject and verb are separated by a clause that provides additional information on the subject. Readers may be more interested in the subject once they know of its contribution to physics.

> *Revised*: Scientists at CERN have announced the detection of the Higgs Boson, which is the final particle proposed by the Standard Model of particle physics. CERN is the world's largest particle physics laboratory, funded collaboratively by twenty European countries.

* Gopen G, Swan J. The Science of Scientific Writing. American Scientist. 1990;78:550–558. Available from http://www.unc.edu/~haipeng/teaching/sci.pdf

† Sections 1.3 and 1.4 provide specific recommendations to improve your prose.

‡ A *sentence* is one or more words that express a complete thought. All sentences contain a subject and predicate.

- The *subject* (noun or pronoun) is the person or thing doing the action.
- The *predicate* (verb) is the action or state of the subject.

More information on sentences is presented in section 1.3.

Put appropriate information in the "topic" and "stress" positions

The *topic position* is at the beginning of the sentence. The topic position links the sentence to existing knowledge or information previously presented and contextualizes the new information presented later in the sentence.

The *stress position* is at the end of the sentence. The reader interprets the information in the stress position as important. Put new information that you want to emphasize in the stress position.

> *Original*: Accident reconstruction, whereby the events before and during a motor vehicle accident are determined based on physical evidence from the scene, is an application of applied classical physics.
> *Revised*: An application of applied classical physics is accident reconstruction, whereby the events before and during a motor vehicle accident are determined based on physical evidence from the scene.

In the original sentence, the reader is unsure of the context surrounding "accident reconstruction" and may expect a more profound ending to the sentence. However, the ending is more general. The revised sentence builds from general to specific and from the known to the unknown.

Present information in logical progressions

A series of sentences related to a single idea form a *paragraph*. While there are many ways to arrange the ideas in a paragraph, a common format is to have each sentence link to existing knowledge and add new information, with the important idea presented near the end of the paragraph. (See page 22 for details and alternative paragraph formats.)

> AB. BC. CD. DE.

In the above pseudo-sentences, the first letter corresponds to existing knowledge or information in the topic position; the second letter corresponds to the new information in the stress position. The four sentences would thus make a paragraph.

Scientists often assume the reader is knowledgeable and can fill in the missing information because the information appears trivial and obvious. While trivial and obvious to the scientist, the missing information may not be obvious to the reader, and coherence in the paragraph is lost.

> AB. BC. DE.

In the above pseudo-sentences, C is not used and D is not introduced. Even if the reader does have the knowledge to determine CD, it is taxing and time-consuming to reread a paragraph while trying to deduce what is missing. Readers without this specific knowledge are lost.

> *Original*: Chemical kinetics is the study of the rates of chemical reactions as a function of concentration and temperature. The many different entities present in food complicate the application of kinetics to cooking. One class of chemical reactions, the Maillard reactions, involves the reaction of amino acids with carbohydrates. Maillard reactions produce the browning and much of the flavor in fried foods such as meat, eggs, and toast. Maillard reactions do not occur in boiled and microwaved food.

In the above paragraph, the phrase that introduces concentration and temperature is irrelevant because neither is mentioned later in the paragraph. Additionally, the paragraph does not explain why Maillard reactions do not occur in boiled and microwaved foods.

> *Revised*: Chemical kinetics is the study of the rates of chemical reactions as a function of concentration and temperature. The many different entities present in food complicate the application of kinetics to cooking. One class of chemical reactions, the Maillard reactions, involves the reaction of amino acids with carbohydrates. Above 140 °C, foods with large quantities of these entities undergo *browning*, which is the source of much of the flavor in fried foods such as meat, eggs, and toast. Maillard reactions do not occur in boiled and microwaved food because the cooking temperature never exceeds 100 °C.

> If the writer is placing information where the reader expects it, the reader perceives greater coherence in the document and is increasingly interested and engaged by the work. Importantly, there is a greater chance that the writer will convey the desired information to the reader.

Deviating from the guidelines

The guidelines presented above are just that: guidelines. The occasional deviation from these guidelines can have profound effects: adding emphasis to key concepts, building suspense and anticipation, drawing the reader into your work, and leading to a memorable conclusion — or completely losing the reader. Be careful!

1.2 Punctuation

Punctuation adds structures to prose to clarify meaning. Below are some common and scientific uses of punctuation.* Section 2.4 explains the unique uses of punctuation in citations.

How important is punctuation? Very! Poor punctuation makes your work difficult, if not impossible, to read. Readers will misinterpret your work. Educated people will question your intelligence and education.

Commas (,) indicate a separation of ideas or elements within a sentence. Use a comma

- to separate dependent clauses†

  Before starting the experiment, ensure that the detector has cooled to –20 °C.

  The Large Hadron Collider (LHC), which is on the border between France and Switzerland, is the world's highest energy particle accelerator.

- before a coordinating conjunction (*and*, *or*, *but*, *for*, *nor*, *yet*, *so*) that joins independent clauses

  There is a linear relationship between absorbance and concentration at low concentration, but the relationship becomes nonlinear at higher concentration.

- to separate items in a series

  This course covers kinetics, equilibria, and thermodynamics.

For purposes of clarity, a comma *is required* before the last item in a series, as illustrated by the following example.

> *An individual intended to give equal portions of his estate to his three children. His will, however, stated "... to my children Jill, Susan and Mark ..." Jill argued that she deserved fifty percent and the other two children deserved twenty-five percent each due to the lack of a comma. Her claim was upheld in court. The court stated that the will should have stated "... to my children Jill, Susan, and Mark ..." for each child to receive one-third of the estate.* — unknown

* Punctuation marks may have additional specialized uses within scientific disciplines.

† *Dependent clauses* are clauses that can be removed without changing the meaning of the remaining sentence. Dependant clauses can be at the beginning of the sentence (introductory) or in the middle of a sentence (interrupting).

Independent clauses are complete sentences, but may be part of a complex sentence.

See section 1.3 for details.

Do not use a comma to separate the digits in a large number. Some countries use a comma as the decimal point. Use a space, preferably a *thin space* (see Table 1.2), to separate the digits and minimize confusion.

299,792,458 m/s → 299 792 458 m/s

Periods (.) are

- placed at the end of a complete thought (sentence)
- in abbreviations to indicate that a portion of the word(s) has been omitted

  Ph.D., M.Sc., Dr.

  For abbreviations that are all capitals or spoken as words, omit the periods.

  NASA, NORAD, CERN

- used as the decimal point (The period is preferred to the comma, but be aware that some European countries use a comma as the decimal point.)

  The sample was irradiated with 337.1 nm radiation from a nitrogen laser.

Semicolons (;) are used to

- join complete sentences to emphasize a close relationship between the sentences, but only if a conjunction is not used to join the sentences

  There are many idioms and clichés in the English language; they should not be used in scientific prose.

- separate items in a series when those items contain internal punctuation

  The active voice is closer to real speech and closer to natural thought processes; produces shorter, concise sentences; and makes arguments more persuasive.

Colons (:) are used to

- join two independent but closely linked clauses, especially if the second clause either proves, explains, or illustrates the first. The first word after the colon may be capitalized at your discretion, but be consistent throughout the work.

  Endeavoring to wow a public audience with highly technical language will have the opposite effect: the audience will not understand what you have written and will have an increased disdain for you and science.

- introduce a list or definition, but only if the preceding clause is a complete sentence.

  This course covers three topics: kinetics, equilibria, and thermodynamics.

- separate components of a fraction or ratio

  Black powder is a 10:15:75 mixture by mass of sulfur, charcoal, and saltpeter.

- separate the hours:minutes:seconds of time

  The solar flare was detected at 03:48:51 UT on 18 July 2013.

Apostrophes (’) are used to

- indicate possession
  - singular nouns: add (’s)

    The scientist’s prose is impeccable.
    The virus’s mode of infection involves ….
    The EPA’s report on airborne particulates …. (the acronym is treated as a word)

  - plural nouns not ending in (s): add (’s)

    The mice’s cage was maintained at 18 °C and 40 percent humidity.

  - plural nouns ending in (s): add (’)

    The students’ success came because of their hard work and dedication.
    The emissions’ carbon monoxide content exceeded 15 ppm by volume.

- indicate the omission of one or more letters from a word to form a contraction (not commonly used in formal writing)

  I will → I’ll　　　　do not → don’t

Apostrophes are often misused. The following examples are all correct and illustrate the different applications of apostrophes.

The scientist’s laboratory.	(A single laboratory managed by one scientist.)
The scientist’s laboratories.	(Multiple laboratories managed by one scientist.)
The scientists’ laboratory.	(A single laboratory managed by multiple scientists.)

Quotation marks (“” , ‘ ’) indicate that the text contained within is a direct quotation. Punctuation associated with the quotation is retained within the quotation marks. When quoting within a quotation, single quotation marks surround the inner quotation. When quoting three or more lines of text, write the quoted text as a separate paragraph indented from the normal text.

Quotation marks can also indicate the application of a different meaning to a phrase: sarcasm, doubt, cynicism, etc.

Question marks (?) indicate that the statement should be read as an interrogative sentence (a question). The interrogative phrase typically contains *who*, *what*, *where*, *when*, *why*, or *how*. Question marks can also indicate uncertainty in a value or statement. In the latter use, the question mark may be enclosed in parentheses: (?).

The structure was built to code, so why did it collapse?

Exclamation marks (!) indicate extreme emphasis — beyond that available within the words themselves — on a statement. Exclamation marks are *rarely* used in scientific work! A maximum of one sentence in a paragraph should receive an exclamation mark.

CAUTION: warming the product above 60 °C causes an explosive decomposition!

Brackets (() , [] , <> , {})

- curved brackets (( ) ; also called *parentheses*)
  - contain supplementary information. The text in parentheses may be omitted without affecting the conveyed information. Commas are also used to include supplementary information; it is at your discretion whether to use commas or parentheses, but the use must be consistent throughout the work.

    Ethylenediaminetetraacetic acid (EDTA) removes specific metal ions.
    Ethylenediaminetetraacetic acid, EDTA, removes specific metal ions.

  - are used as a shorthand to denote either singular or plural

    Once the sample(s) is (are) analyzed, …

- square brackets ([ ])
  - are used to insert information into direct quotes

    The BBC News Service reported, "A NASA study has found that the continent [Antarctica] is losing around 152 cubic km of ice each year."

  - represent the concentration of a solution in moles per liter
  - contain citation information in some citation styles
- angle brackets (< >)
  - are used to contain information that must be inserted by the reader

    <insert example here>

  - are used in mathematics to denote less than, greater than, and other advanced mathematical operations
- curly brackets ({ } ; also called *braces*) enclose the elements in a set

  The first five primes are $\{2,3,5,7,11\}$.

- In mathematics, brackets indicate the order of operations.

  $5+4\cdot 3^2=41$ $\quad (5+4)\cdot 3^2=81$ $\quad \left[(5+4)\cdot 3\right]^2=729$

Ellipses (…) are used to indicate that text was omitted from a quotation. An ellipsis can also indicate an intentionally unfinished sentence or list. When using an ellipsis, ensure the remaining text retains the original meaning and conveys the desired information.

Style guides differ on whether there should be spaces around ellipses. Whatever you choose, be consistent!

Dashes (- , – , —)

- The hyphen (-) is used to
  - join compound words

    half-life   non-reactive

  - join compound modifiers preceding a noun, which imparts a different meaning than the separate words

    high-frequency noise   second-year mathematics
    mid-Atlantic trench   Bose-Einstein condensate

  - join words that spell numbers

    Approximately one-quarter of the samples are contaminated.
    Twenty-eight monitoring stations were set up throughout the forest.

  - represent a negative number
  - hyphenate words at their syllables for spacing within a document
- An en dash (– , about the width of the letter N) is used to indicate numerical ranges, replacing the word *to.*

  Samples 3 – 7 showed signs of degradation after the freezer failed and the samples were at room temperature for over 24 hours.

- An em dash (—, about the width of the letter M) is used to denote a major break in a sentence. Dashes could be replaced with commas, but the dash adds greater emphasis to the separation. The portion separated by dashes could be removed without changing the meaning of the remaining sentence.

  I would appreciate recommendations — ones that have minimal implementation cost — on improving the sampling procedure.

Style guides differ on whether there should be spaces around the en-dash and em-dash. Whatever you choose, be consistent!

Solidus (slash) (/) is used to indicate fractions and as a substitute for *or* when indicating a choice. There are no spaces between the solidus and the adjoining words or numbers.

The survival rate was $^{46}/_{53}$.   Y/N
(kg m)/s^2 male/female

Formatting text

Formatting draws attention to text. A key factor in formatting text is that it must be done sparingly to maintain its effectiveness.

Italics are used to emphasize a word or phrase, to introduce new terms, and to identify foreign words not common in English. For example, biologists italicize the Latin names of organisms.

Bold and ***bold-italics*** have no defined use. Headings are commonly bold. (*Communicating Science* uses bold-italics to highlight a word being defined in the sentence.)

However you choose to use formatting in your work, be consistent and use formatting sparingly.

Additional characters

In addition to the obvious keys on the keyboard, computer fonts (Arial, Helvetica, Times New Roman, Calibri, ...) contain hundreds to thousands of additional characters. Most word processing programs have a Character Map to view these characters and insert them into your document. Each character also has a unique Unicode sequence that allows you to insert them without using the Character Map. How this is done depends on the program and the operating system (consult the Help menu for instructions). Tables 1.1 and 1.2 present some of the additional characters commonly used in science.

STIX fonts

A consortium of scientific publishers is developing a single comprehensive font set that contains all the fonts used by the scientific and engineering community. This project will aid writers and publishers in preparing and publishing scientific documents. Dubbed the Scientific and Technical Information Exchange (STIX) project, the STIX fonts are available from http://www.stixfonts.org.

As of 2013, the STIX fonts are not fully functional in Microsoft Office® products.

Table 1.1 The characters of the Greek alphabet.

Greek alphabet					
name	**case: lower**	**upper**	**name**	**case: lower**	**upper**
alpha	α	Α	nu	ν	Ν
beta	β	Β	xi	ξ	Ξ
gamma	γ	Γ	omicron	ο	Ο
delta	δ	Δ	pi	π	Π
epsilon*	ε, ϵ	Ε	rho	ρ	Ρ
zeta	ζ	Ζ	sigma	σ	Σ
eta	η	Η	tau	τ	Τ
theta	θ	Θ	upsilon	υ	Υ
iota	ι	Ι	phi*	ϕ φ	Φ
kappa	κ	Κ	chi	χ	Χ
lambda	λ	Λ	psi	ψ	Ψ
mu	μ	Μ	omega	ω	Ω

* There are two common written forms of epsilon and phi.

Table 1.2 Commonly used characters in Times New Roman and Symbol fonts.

Selected characters					
name	**char.**	**name**	**char.**	**name**	**char.**
cents	¢	plus-minus	±	therefore*	∴
Euro currency	€	multiplication	·	for any*	∀
British currency	£	multiplication	×	there exists*	∃
open bullet	◦	division	÷	contains*	∈
small bullet	•	tensor product*	⊗	does not contain*	∉
large bullet	●	approx. equals	≈	intersection*	∩
copyright	©	not equals	≠	union*	∪
registered	®	identical	≡	superset*	⊃
trademark	™	less than or equal	≤	subset*	⊂
degree	°	greater than or equal	≥	not a subset*	⊄
angle*	∠	prime	′	empty set*	∅
perpendicular*	⊥	double prime	″	dagger	†
arrows	←↑↓→	proportional*	∝	double dagger	‡
arrows	⇐⇑⇓⇒	infinity	∞	male, female	♂, ♀

* These characters are from the Symbol font.

1.3 Document structure

Section 1.1 provides general strategies for producing quality scientific prose. The following sections provide specific suggestions for improving your prose.

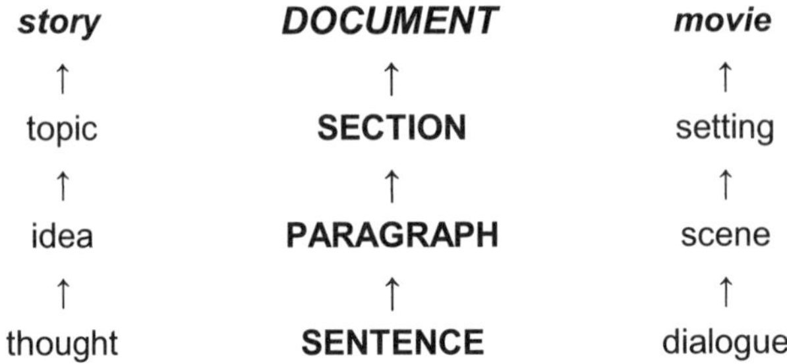

Figure 1.5 A comparison of the building blocks of a story, academic work, and movie.

Sentences

Sentences are the building blocks of prose, with every sentence presenting a single thought. The basic elements of a sentence are the

- ***subject*** (***noun*** or ***pronoun***): the person or thing doing the action
- ***predicate*** (***verb*** or ***verb phrase***): the action or state of the subject
- ***object***: the person or thing receiving the action

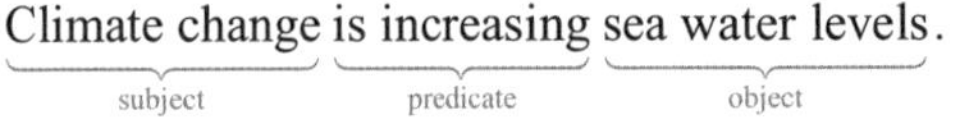

Some sentences omit either the subject or the object if it is obvious what or who is receiving the action. However, most scientific sentences include them for clarity.

Complements are elements that provide additional information about the subject, predicate, or object.

We mixed the chemicals in the beaker with a stir rod.
subject predicate object object complement predicate complement

Beyond these elements, sentences are a collection of independent and dependent clauses. An ***independent clause*** can stand as a sentence, but can also be a component of a compound or complex sentence. A ***dependent*** (***subordinate***) ***clause*** modifies another clause but is itself not a complete sentence.

At 60 °C, the solution turned blue, indicating that a reaction was occurring.
dependent clause independent clause dependent clause

Simple sentences are independent clauses.

Compound sentences are two or more independent clauses joined with a *conjunction*: and, or, but, for, nor, yet, so, ….

Complex sentences contain one or more dependent clauses.

Simple: We mixed the chemicals in the beaker. (independent clause) The beaker got warm. (independent clause)

Compound: We mixed the chemicals in the beaker, (independent clause) and the beaker got warm. (independent clause)

Complex: After we mixed the chemicals in the beaker, (dependent clause) the beaker got warm. (independent clause)

Short sentences — fewer than ten words in length — are simple and clear, but multiple short sentences result in choppy, abrupt text that is difficult to read. Compound and complex sentences — up to around thirty words in length — are useful for conveying connected information. Sentences over thirty words are difficult to understand and exceedingly difficult to write in a way that accurately and unambiguously conveys information. A good practice is to vary sentence length.

Original: Water quality declined in July. The decline occurred because of heavy rainfall. The rainwater washed sediment into the reservoir. The sediment overloaded the water treatment system. Normal water quality was restored in mid-August.
Revised: Water quality declined in July because heavy rainfall washed sediment into the reservoir, overloading the water treatment system. Normal water quality was restored in mid-August.

Dependent clauses can be restrictive or nonrestrictive.

Restrictive clauses limit or modify the noun in the sentence. Removing the clause changes the meaning of the sentence. Restrictive clauses are not separated from the independent clause by punctuation.

The lab tested the samples (independent clause) that were less than 15 days old. (restrictive dependent clause)

This sentence implies that some samples were more than 15 days old.

Nonrestrictive clauses describe or provide additional detail on the noun. Removing the clause removes information but does not change the meaning of the sentence. Nonrestrictive clauses are often separated from the independent clause by a comma.

The lab tested the samples, (independent clause) which were less than 15 days old. (nonrestrictive dependent clause)

This sentence implies that all samples were less than 15 days old.

Recall from section 1.1 that you should put information where the reader expects it:

- The first piece of information in a sentence (in the *topic* position) should link the sentence to information already presented and contextualize the new information presented later in the sentence.
- The last piece of information in a sentence (in the *stress* position) contains new information the writer wants to emphasize.

Paragraphs

A series of sentences related to a single idea forms a *paragraph*. There are usually between four and eight sentences per paragraph. One sentence in the paragraph — the ***topic sentence*** — expresses the essential idea of the paragraph, with the remaining sentences supporting or illustrating the topic sentence or linking it with the remainder of the document. The topic sentence can be anywhere in the paragraph.

One paragraph format has each sentence linking to existing knowledge and adding new information, with the topic sentence near the end of the paragraph. An alternative paragraph format has the topic sentence as the first sentence. The remainder of the paragraph then adds information to support that idea and place it in context of the document.

Experienced writers use both paragraph styles, choosing the style that best conveys information in a clear, concise, and precise manner and maintains cohesion with the remainder of the document. For example, in each of the following sentences, identify the clauses in the topic and stress positions, arrange the sentences into a coherent paragraph, and identify the topic sentence of the paragraph.

1: The time intervals are approximately constant because the rates of tectonic plates movement and strain accumulation along the fault are roughly uniform.

2: For example, the recurrence intervals along the southern segment of the San Andreas fault are 145 years with variations of several decades.

3: Major earthquakes along a given fault line do not occur at random intervals because it takes time to accumulate the strain energy for the rupture.

4: Indeed, the time interval of major earthquakes along fault lines often varies by a factor of two.

5: However, the time intervals will vary depending on the amount of strain required to start the earthquake and the amount of strain released during each earthquake.

The reconstructed paragraph is on page 58.

Paragraphs should link to the preceding paragraphs and lead into the subsequent paragraph, providing continuity and coherence to the information being presented.

Getting the reader to follow your logic is challenging, but critical to producing a clear and coherent document. Simply stating fact after fact does not interest, engage, or retain the reader. To do these things, you must

- properly structure sentences and paragraphs
- properly punctuate compound and complex sentences
- use advanced punctuation — colons, semi-colons, dashes — to inform the reader of connections between information
- write at the level of the reader: start with their existing knowledge and build their understanding

To help the reader follow your logic, use words that indicate relationships between ideas. Table 1.3 presents a selection of relationships and words that build relationships.

Table 1.3 Words used to build relationships between thoughts and ideas in a work.

Purpose	Selected transition words
add supporting ideas and information	additionally, again, also, as well, equally important, furthermore, in fact, in addition, indeed, moreover, similarly
identify conditions	considering, for the most part, generally, granted, if, occasionally, often, ordinarily, rarely, sometimes, usually
introduce opposing ideas and information	although, but, conversely, despite, however, in contrast, nevertheless, on the contrary, still, whereas, while, yet
show time, sequence, and causation	first, second, before, after, as a result, because, consequently, finally, hence, next, since, subsequently, therefore, then, thus
provide emphasis	certainly, essentially, in fact, importantly, in particular, notably, particularly, primarily, significantly, specifically
provide examples	consider, for example, for instance, in this case, look at, specifically, such as, to demonstrate, to illustrate
summarize and conclude	finally, in brief, in conclusion, in review, in summary, simply, that is, therefore, thus, to conclude, to sum up, ultimately

Sections

Sentences flow logically from one to the next until an entire idea is presented in a *paragraph*. Paragraphs flow logically from one to the next until an entire topic is presented in a *section*. However, there can be significant discontinuity between sections.

In laboratory reports, common sections include the *Introduction*, *Materials*, *Procedure*, *Results*, *Discussion*, and *Conclusion*. These sections present important, but distinct, information that together forms the complete report. In *Communicating Science*, the chapters are sections and the numbered headings are subsections within the chapters. These subsections present distinct information that together conveys the complete topic. The chapters present a complete introductory guide to communicating science.

Section headings

Every section starts with a heading. Traditional headings are as stated above: *Introduction*, *Materials*, …, *Conclusion*. These headings are general, providing little information on the contents. Scientific writing is currently transitioning towards headings that are specific, substantive, and that provide information on the section contents.

Procedure → Setup and operation of the ____
Results and discussion → Analysis of the ____ data

While I recommend that you use specific, meaningful headings in your documents, you must conform to the requirements of instructors, employers, and publishers in work you prepare for them. You may find that you have more freedom in using descriptive headings when preparing project reports, essays, posters, and oral presentations, but have less freedom when preparing laboratory reports and theses.

Documents

Chapter 5 explores how sentences, paragraphs, and sections combine to create documents and presentations for scientific and non-scientific audiences.

1.4 Document language

This subsection presents strategies to prepare interesting and engaging scientific prose. In the process, reasons why scientific prose is perceived as abstract, vague, difficult, and tedious to read are identified.

Scientific terms

Many words have defined scientific meanings that differ from common usage. Only use these words in their scientific context when preparing scientific works.

abstract: a summary of a scientific work
aerosol: a small particle dispersed in air (for example, smoke and fog)
bias: a systematic deviation (statistics); constant current or voltage offset (electronics); a preference or prejudice towards a certain result (academic integrity)
confidence level: a measure of the reliability of a result
data: scientific observations that have been collected and recorded
error: a difference between the measured and expected value; range of acceptable values at a certain confidence level (also *uncertainty*)
fact: an observation that has been repeatedly confirmed and is accepted as true
hypothesis: a proposed and testable explanation of observed phenomena
investigation: an exploration of existing knowledge
law: a concise statement that describes the relationship between phenomena
mass: the amount of matter in a body
mean: a calculation of the average
model: a simplified representation of a complex system; a simulation
opinion: a belief based on scientific reasoning
organic: a carbon-containing molecular entity
proof: evidence and results that support a hypothesis (experimental science); the derivation of a mathematical equation or logical statement (mathematics)
radiation: energy in the form of high-energy subatomic particles or electromagnetic radiation
research: an exploration of the unknown
significant: supported by statistical analysis
solution: a homogeneous mixture
state: the properties of matter in a system: temperature, pressure, physical form
theory: a set of rigorously tested statements or principles that explain and are used to make predictions about a phenomena; an explanation of observed phenomena that has been confirmed through scientific testing and is broadly accepted by the scientific community as true
tolerance: range of acceptable deviation
uncertainty: range of acceptable values at a certain confidence level (also *error*)
weight: a force on a body calculated as *mass* × *acceleration*

Grammatical tenses, aspects, and moods

Most native English speakers write reasonable sentences without understanding the grammatical foundations of sentences. This section introduces grammatical tenses, aspects, and moods to build these foundations and to help others whose first language is not English.

There are three tenses and four aspects in the English language. The ***tense*** indicates when an action occurred. The ***aspect*** indicates how that action relates to the flow of time.

- *simple*: encodes no information about time.
- *perfect*: indicates that the action is complete.
- *progressive*: indicates that the action is continuing.
- *perfect progressive*: indicates that the action is continuing but with a defined completion point.

Table 1.4 Grammatical tenses and aspects in the English language.

tenses	aspects: simple	perfect	progressive	perfect progressive
Past	wrote	had written	was/were writing	had been writing
Present	write	has/have written	am/are/is writing	has/have been writing
Future	will write	will have written	will be writing	will have been writing

The ***mood*** indicates the type of sentence. Some common English moods are presented below:

- *indicative*: presents factual information.

  The fossils were discovered after the flood waters receded.

- *conditional*: presents an action that is dependent upon other factors.

  The walkway might collapse if the steel strength is not up to specifications.

- *imperative*: expresses commands, instructions, or direct requests.

  Let's attempt this experiment. Get the fire extinguisher!

- *interrogative*: asks a question.

  What is causing the vibrations at high velocities?

- *subjunctive*: presents a wishful, hypothetical, or suggestive action. The use of the subjunctive mood is not common in written scientific work.

  It is important that you be at the presentation.

Grammatical tenses in science writing

If something is true today, write it in the present tense. In scientific documents, much of the *Introduction* and the *Data*, *Results*, *Discussion*, and *Conclusion* should be written in the present tense.

This report investigates ....
Figure 3 shows ....

If something occurred and is complete, write it in the past tense. In scientific documents, much of the *Methods* or *Procedure* should be written in the past tense.

Einstein's Special Theory of Relativity, which he published in 1905, proposed that the speed of light is a constant independent of the reference frame of the observer.

- "published in 1905" — past tense — done and completed
- "the speed of light is a constant" — present tense — correct then and today

The speed of light *is* exactly 299 792 458 m/s.
The speed of light *was* defined as an exact value by the CGPM in 1983.

If something should or will happen, write it in the future tense. In scientific documents, some of the *Discussion*, *Conclusion*, and *Future studies* sections should be written in the future tense.

Future experiments will test this hypothesis.

Active and passive voice

The *voice* is the relationship between a subject and verb.

- *active voice*: the subject is doing the action. The focus is the subject.

  Students recorded data. Polly wants a cracker.

- *passive voice*: the subject receives the action of the verb. The focus is the action.

  Data was recorded by students. A cracker is wanted by Polly.

As seen from the above examples, the active voice is more concise and usually preferred in writing. However, there is a tendency amongst scientists to write in the passive voice. This tendency is not a necessity of scientific writing, but has been the culture amongst scientists and expected or required of science students, thus perpetuating the culture. Thankfully, *this tendency is decreasing!* More and more scientists are choosing to write in the active voice to make their work clear, concise, and more interesting to the reader.

In many cases, switching between the active and passive voice involves reversing the subject and object or subject and verb.

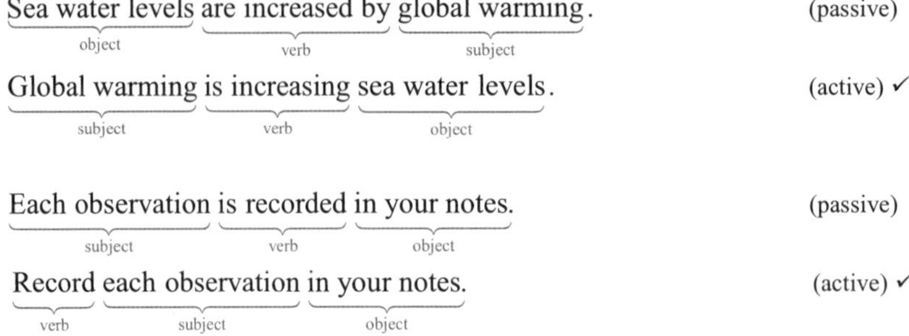

Example: given the passive sentence "You are loved by me", what is the equivalent active sentence?

Phrases are also active or passive:

enthalpy of reaction (passive)
reaction enthalpy (active) ✓

The active voice is preferred in most prose. The active voice is engaging, easier to read, and draws the reader into the prose. The active voice best reflects real speech and natural thought processes; produces shorter, concise sentences; and makes arguments more persuasive.

In science, the passive voice was dominant because it presented the science as independent of the scientist. Demanding that students write in the passive voice addressed the problem of over-use of the first-person in the active voice.* However, in the opinion of a growing number of scientists and most academic journals, the active voice is acceptable and valuable in scientific prose. Indeed, numerous prominent scientists — Curie, Darwin, Einstein, Faraday, Feynman, and Watson and Crick — regularly used the active voice in their scientific publications. For example, in their seminal article proposing the structure of DNA, Watson and Crick state,

> *We wish to suggest a structure for the salt of the deoxyribose nucleic acid ....*
> Nature. 1953;171:737–738.

The passive voice is a major reason why scientific prose is difficult to read, thus perpetuating the myth that science is boring and unintelligible.

* First-person is not a requirement when writing in the active voice (see page 30).

In general, the passive voice increases confusion, increases text length, and makes writing less lively.

An increase in desertification is predicted by the model. (passive)
The model predicts an increase in desertification. (active) ✓
The model predicts increased desertification. (active) ✓

Essays are written by students. (passive)
Students write essays. (active) ✓

However, there are times when the passive voice is preferred. Choosing the active or passive voice depends on whether the emphasis is on the subject (active) or the object (passive). Use the passive voice

- to put emphasis on the object and action, not the subject committing the action. This is common in *Methods* and *Procedure* sections of scientific documents.*

  The samples were heated to 65 °C for 24 hours.
  The glassware was sterilized in an autoclave.

- when the subject committing the action is unknown or unimportant.

  The equipment was damaged yesterday.
  The stock solution was contaminated.

Unfortunately, you may have an instructor, employer, or publisher who insists that you write in the passive voice. Call them an *old fogy*, but do what they say.

General revisions that change from passive to active voice include

are/have/is ____ed → ____ (are mixed → mix; is heated → heat)
x of *y* → *y x* (enthalpy of reaction → reaction enthalpy)
x of a *y* → *y x* (activity of a solute → solute activity)
was/were used to → <delete> (the laser ~~was used to~~ ionize → the laser ionized)

Below are some additional examples of converting from the passive to active voice.

can be → is/are
has been → was/were
have been → were/are
in which/for which → where
is a measure of → measures
is capable of → can
is used to detect → detects
it is assumed → I/we assumed
the result of this is seen as the deviation → the result is the deviation
there were four samples that were analyzed → we analyzed four samples

* Instructions in a manual are written in the active voice (see section 5.11).

Heat the samples to 65 °C for 24 hours.
Sterilize the glassware in an autoclave.

First, second, and third person

When examining the relationship between the author and the audience, we find that three relationships exist:

- *first person*: the author refers to themselves.

  I conducted a finite element analysis of the bridge.

- *second person*: the author refers to the reader.

  You conducted a finite element analysis of the bridge.

- *third person*: the author refers to someone or something other than the author or the reader.

  They conducted a finite element analysis of the bridge.
  Winthrop Engineering conducted a finite element analysis of the bridge.

Table 1.5 lists pronouns used in communicating in the first, second, and third persons. In addition to the third person pronouns, the noun can be explicitly stated as illustrated in the example above, where the third person is "Winthrop Engineering".

Table 1.5 Pronouns used to communicate in the first, second, and third person.

person:	first	second	third
singular	I, me, my, mine	you, your(s)	he, him, his, she, her(s), it(s), they,* them,* their*
plural	we, us, our(s)	you, your(s)	they, them, their(s)

* *They*, *them*, and *their* are increasingly being used as a singular gender-neutral pronouns.

A common objection to using the active voice is that it leads to increased use of the first person (I/we) in prose. So? A scientist proposed the research, designed and conducted experiments, and analyzed the data. To not identify the actions of the scientist is not logical. It is entirely acceptable to use the first person when describing actions where it is reasonable to emphasize the person's involvement.

The work of Smith found ____, but this investigation found .... (active, third person)
Smith reported ____, but we found .... (active, first person) ✓

This study investigates the .... (active, third person)
I/we chose to study the .... (active, first person) ✓

The ore samples were crushed in a ball mill. (passive, third person)
A ball mill crushed the ore samples. (active, third person) ✓
We crushed the ore samples in a ball mill. (active, first person) ✓

However, the first and second person can be overused in academic prose. Overuse by new writers occurs when writers are uncomfortable with scientific terminology and/or academic writing. First person is overused when multiple sentences contain *I* or *we*, resulting in prose that lacks a professional tone.

> *Original*: After confirming the extraction process worked in the laboratory, we were commissioned to build a pilot plant and scale-up the process. I developed the feed and thermal control systems, while Tory designed the monitoring systems and chemical control systems. During the initial pilot tests, Tory observed that the product was not pure, indicating that the extraction was not 100 percent efficient. We are currently trying to determine why this is occurring.
>
> *Revised*: After confirming the extraction process worked in the laboratory, we were commissioned to build a pilot plant and scale-up the process. Feed and chemical control, thermal control, and internal monitoring systems were developed for the plant. Initial tests found contamination of the product stream, and we are currently determining the source of the contamination.

Second person is used minimally since academic works typically inform, not direct, the reader.* One exception is instructions and procedures. However, second person references can often be removed or converted to third person without affecting the information being conveyed.

When tagging salmon, you should know that	(second person)
When tagging salmon, one should know that	(third person)
Before you enter the laboratory, ensure that your laboratory coat and safety glasses are completely on.	(second person)
Everyone must ensure their laboratory coats and safety glasses are completely on before entering the laboratory.	(third person)

Many who argue against the use of active voice and first person argue that the passive voice is objective and impartial. Writing does not make a scientist objective and impartial. A scientist achieves objectivity and impartiality by conducting impartial experiments and completing an unbiased analysis of the data.

> *Scientific misconduct* is the falsification or improper interpretation of data. One argument for writing in the passive voice is that it removes the scientist from the research. However, a review of retracted research articles shows that many are written in the passive voice.

* *Communicating Science* contains substantial second person prose, which engages and increases your involvement. Realize that one goal of *Communicating Science* is to get you engaged so that you practice and learn these skills.

Nominalizations

A ***nominalization*** is the creation of a noun from a verb or adjective. Nominalizations focus the prose on objects and concepts and away from the action and person committing the action. Because of this, nominalizations tend to complement the passive voice. Additionally, nominalization of an important verb obscures the information the writer is trying to convey, which decreases clarity.

Nominalizations are formed by adding a suffix to the verb or adjective. Common suffixes include *–ion*, *–ation*, *–ing*, *–ity*, *–ment*, and *–ance*. Another type of nominalization occurs when the same word is both a verb and noun. Table 1.6 lists common science nominalizations.

Table 1.6 Nominalization of verbs and adjectives common in scientific prose.

Verb	Nominalization	Verb	Nominalization
analyze	analysis	observe	observation
compare	comparison	occur	occurrence
consider	consideration	perform	performance
deduce	deduction	prove	proof
discover	discovery	react	reaction
dissect	dissection	regulate	regulation
examine	examination	resist	resistance
experiment	experiment	report	reporting
fail	failure	understand	understanding
grew	growth	vary	variance
inject	injection	**Adjective**	**Nominalization**
increase	increase	applicable	applicability
investigate	investigation	bright	brightness
know	knowledge	different	difference
legalize	legalization	difficult	difficulty
measure	measurement	intense	intensity
move	movement	weak	weakness

In the following examples, the verb is italicized and the nominalization is underlined.

An investigation *was conducted* into the friction coefficient of sandpaper. (nominalized, passive)
The friction coefficient of sandpaper *was investigated.* (passive)
We *investigated* the friction coefficient of sandpaper. (active) ✓

In the first sentence, the important action verb, *to investigate*, is nominalized to the noun *investigation* and the sentence verb becomes the passive *was conducted.* The second sentence is an improvement by using the correct action: *investigate.* However, this sentence is written in the passive voice and does not indicate the subject conducting the action. The third sentence, written in the active voice, presents both the actors and the action.

Considering the above sentences,

- Which was easier to understand on the first reading?
- Which is closest to spoken language?

For most people, the third sentence has the greatest clarity and is closest to spoken language. The third sentence is also the most concise, requiring three fewer words than the first sentence to convey the same information. Having the greatest clarity and concision, the third sentence should characterize your writing.

Nominalizations create a disconnect between structure and meaning: the action is not found in the verb, where readers expect it to be.

We *performed* an analysis of the data. (nominalized)
We *analyzed* the data. (active) ✓

In the first sentence, the verb is *performed*, but the intended message is that the data was analyzed.

Overuse of nominalizations is another major reason why scientific prose is difficult to read and challenging to comprehend. Indeed, the passive voice and nominalizations together are the two largest contributors to impenetrable scientific prose (and legal and business prose). Observe how the active form is closer to natural speech, clearer, and more concise in the following examples.

The engineers *conducted* an investigation of the accident. (nominalized)
The engineers *investigated* the accident. (active) ✓

The ground state geometry optimization *was done* using DFT. (nominalized, passive)
The ground state geometry *was optimized* using DFT. (passive) ✓

Reaction product identification *was conducted* using NMR spectrometry. (nominalized)
The reaction products *were identified* using NMR spectrometry. (passive)
I *identified* the reaction products using NMR spectrometry. (active) ✓

The archaeologists *made* a thorough documentation of the ruins. (nominalized)
The archaeologists thoroughly *documented* the ruins. (active) ✓

In a few contexts, nominalizations do increase clarity and concision.

- Nominalizations function as transition words, linking successive sentences together. Table 1.3 lists many transition words; some are nominalizations.

  This observation *is consistent* with our hypothesis. (nominalized) ✓
  We *observe* that this analysis is *consistent* with our hypothesis. (active)

- Nominalization of verbs that are not the main action verb better conveys the desired information.

  I do not *understand* their intentions. (nominalized) ✓
  I do not *understand* what they *intend*. (active)

  High-intensity farming has *increased* the erosion rate and *decreased* the quality of the remaining soil. (nominalized) ✓
  High-intensity farming has *increased* the rate soil *is eroded* and *decreased* the quality of the remaining soil. (active)

  Erosion, deforestation, desertification, and climate change all contribute to *decreasing* global food supplies. (nominalized) ✓
  <Active version not reasonably possible.>

- Nominalizations at the end of a sentence provide closure.

  Fertilizer *accelerates* plant growth. (nominalized) ✓
  Fertilizer helps plants *grow* faster. (active)

- Some nominalizations have been accepted as common nouns.

  teacher reaction
  teaching deviation

Analogies, similes, and metaphors

An ***analogy*** is a comparison made to illustrate the similarities between two things. Analogies are commonly used to draw a connection between something known to the reader and something new, which is valuable when presenting new concepts.

Heme is the oxygen binding site in haemoglobin. Heme analogues in myoglobin and fetal haemoglobin also bind oxygen, but with a different affinity to facilitate oxygen transport. Heme analogues in other proteins facilitate reduction:oxidation reactions, such as the production of ATP in cytochromes.

Nerve signal propagation has been compared to electric currents and the brain has been compared to a computer.

Our understanding of the atom has evolved through analogies: billiard ball, plum pudding, beehive, solar system, and finally to a quantum mechanical model.

A ***simile*** is a figure of speech where two things are compared explicitly using terms such as *like* or *as*. These terms indicate that there are limitations in the comparison.

The eye is like a video camera.
Carbon fiber fabric is as strong as fiberglass.

A ***metaphor*** is a figure of speech where one thing *is* a second thing. Metaphors are a stronger comparison than similes.

Evolution is a game of chance.
The fog was a thick soup.

In science, analogies and metaphors are commonly used to present new concepts in terms of the reader's existing knowledge. Scientific models are themselves analogies of a phenomena.

Conversely, poor analogies introduce misconceptions and create confusion rather than understanding. Readers without the necessary background will be confused, and many metaphors have become clichés (see page 53).

The comparison of water waves to explain light waves led to the incorrect hypothesis of the *aether*, a medium for the transmission of light.

Reading the statement "The experiment worked as well as a lead balloon" introduces two possible meanings: when "lead" is interpreted as *forefront*, the reader is confused by the sentence; when "lead" is interpreted as the heavy metal, the reader perceives that the experiment did not work.

Metaphors comparing the body, world, and universe to a machine necessitated the existence of a creator to build the machines.*

You do not really understand something unless you can explain it to your grandmother. — Einstein

* Goonay Z. Metaphors in Science. Fountain Magazine [internet]. 2002 [cited 01 November 2013];40. Available from http://www.fountainmagazine.com/Issue/detail/Metaphors-in-Science

Split infinitives

An ***infinitive*** is a phrase composed of the word *to* and a verb.

A ***split infinitive*** has an adverb between *to* and its accompanying verb. The historical rule, based on applying Latin rules to English, has been to not split infinitives. However, this rule is slowly disappearing. Indeed, split infinitives often produce text with greater clarity, coherence, and precision. Split infinitives can also produce text that is more confusing.

The opening sequence to Star Trek contains a well known split infinitive. Placing *boldly* in any other location does not convey the same meaning.

…to boldly go where no one has gone before. (split infinitive) ✓
…to go boldly where no onc has gone before. ✗

We need to really focus on completing the assignment. (split infinitive) ✓
We really need to focus on completing the assignment. ✓

It is difficult to rewrite the following sentences without a split infinitive.

The goal of scientific communication is to unambiguously convey accurate and detailed technical information to others.
The new equipment will more than triple the analysis rate.
The physician asked the patient to gradually decrease their medication.

Subject-verb agreement

There must be agreement between the number of subjects and the singular/plural form of the verb.

The first sample was … Three samples were …
High concentrations of amino acids were … A high concentration of lysine was …

To test for subject-verb agreement, remove all the phrases that separate the subject and verb. With the subject adjacent to the verb, confirm the subject-verb agreement. Note that it is better to keep the subject and verb close together, so consider rewriting the sentence to improve clarity.

If multiple subjects form a singular collective subject, use a singular verb.

Drinking and driving is dangerous. ("Drinking and driving" being one act.)

If the verb applies to multiple separate subjects that are a mix of singular and plural, use a plural verb.

The bridge and the cars were damaged by the collision.
The cars and the bridge were damaged by the collision.

Learning occurs when new experiences and new information integrate into our pre-existing knowledge framework.

Singular and plural Latin and Greek words

Singular Latin words commonly end in *–um*; singular Greek words commonly end in *–on*. The plural of these words ends in *–a*.

bacterium; bacteria	*E. coli* is the preferred bacterium in microbiology.
medium; media	Bacteria were cultured on agar media.
phenomenon; phenomena	An interesting phenomenon occurs when …

In Latin, *datum* is singular and *data* is plural. However, in everyday use, the meaning of the word *data* is changing. Many scientists now use *data* in both singular and plural contexts.

The data are reported in Table 3.	(traditional)
The data is reported in Table 3.	(modern)
This datum does not fit the linear trend.	(traditional)
This data point does not fit the linear trend.	(modern)

From another perspective, the word *data* is becoming like *information*. Information is singular but may represent one fact or many facts.

Data is not the only word that has split from its roots:

- *Agenda* (historically plural) now refers to a single list of things to do. *Agendum* is rarely used and *agendas* is commonly used.
- *Criteria* (historically plural) now refers to both one and multiple standards. *Criterion* is still used, but with decreasing frequency.
- *Stadium* is a single sports center. *Stadiums* is used when referring to multiple sports centers. *Stadia* is rarely used.

Parallel grammatical structure

When presenting multiple pieces of information, using the same grammatical structure — *parallel structure* — allows the reader to recognize the separate pieces.

Writing in science is not the same as literature writing.	(error)
Writing in science is not the same as writing in literature.	(parallel) ✓
Practice reviewing and editing your work and the work of others.	(error)
Practice reviewing and editing your work and others' work.	(parallel) ✓
Experiments were conducted in methanol and water.	(error)
Experiments were conducted in methanol and in water.	(parallel) ✓
The bacteria were tested for their resistance to pH, salinity, and heat.	(error)
The bacteria were tested for their resistance to pH, salinity, and temperature.	(parallel) ✓

Lists

Presenting information in list form emphasizes the information in the list and is easier to read and follow. If the order of information matters, as in instructions, use a numbered list; otherwise, use a bulleted list. The order of items in a bulleted list should follow a logical progression (preferred), be listed alphabetically, or be in order of increasing line length.

Materials
- 500 g sodium carbonate
- 1.0 L concentrated acetic acid
- 1 box of gloves, medium

Procedure
1. Go directly to chemistry stores.
2. Obtain required chemicals and equipment.
3. Return directly to the laboratory.

When preparing lists:
- Present the information in the list in parallel structure.
- Omit punctuation in bulleted text unless the text forms one or more complete sentences.
- Do not capitalize the first letter unless it begins a complete sentence or is a proper noun.

Abbreviations, acronyms, and initialisms

An *abbreviation* is a contracted form of a word.

Doctor → Dr.
Master of Science → M.Sc.
Street → St.
Doctor of Philosophy → Ph.D.

An *acronym* is a word formed from the initial letters of the words in a proper noun. While the letters abbreviate several words, the periods are often removed in the acronym.

NASA	North American Space Administration
UNESCO	United Nations Education, Scientific, and Cultural Organization

Some acronyms have become so commonly used that they have become common nouns and are not capitalized.

laser	light amplification by the stimulated emission of radiation
radar	radio detection and ranging
scuba	self-contained underwater breathing apparatus

An *initialism* is a set of letters formed from the initial letters of the words in a proper noun. While an acronym is a spoken word, each letter in an initialism is pronounced separately.

DVD	digital video disk
ESA	European Space Agency
LIF	laser-induced fluorescence
MS	mass spectrometer
NMR	nuclear magnetic resonance
RF	radio frequency
STM	scanning tunneling microscope
WHO	World Health Organization

Acronyms and initialisms are common in science. Capitalized acronyms and initialisms should be introduced in parentheses or commas when the phrase is first used. After that, the acronym or initialism should be used. Acronyms that are common nouns (such as laser) do not need to be defined.

mass spectrometry (MS)	mass spectrometry, MS,
radio frequency (RF)	radio frequency, RF,
astronomical unit (au)	astronomical unit, au,

Avoid starting a sentence with an abbreviation, acronym, or initialism unless the abbreviation is part of a proper noun.

Dr. Zweitzer gave permission to conduct the experiment.
St. John's Hospital has a leading cardiac care research center.

British vs. American English

The spelling of some words varies between British English and American English as illustrated in Table 1.7. Both spellings are "correct", although writers, editors, and readers may prefer one spelling.

Table 1.7 Examples of the British and American spellings of words.

British	American	British	American	British	American
-re	***-er***	***-our***	***-or***	***-ae-, -oe-***	***-e-, -o-***
centimetre	centimeter	armour	armor	archaeology	archeology
centre	center	behaviour	behavior	gynaecology	gynecology
fibre	fiber	colour	color	leukaemia	leukemia
kilometre	kilometer	favourite	favorite	orthopaedic	orthopedic
litre	liter	flavour	flavor	paediatric	pediatric
metre	meter	neighbour	neighbor	palaeontology	paleontology
-se	***-ze***	***-l-***	***-ll-***	***-ence***	***-ense***
analyse	analyze	enrolment	enrollment	defence	defense
catalyse	catalyze	fulfil	fulfill	licence	license
colonise	colonize	skilful	skillful	offence	offense
emphasise	emphasize	***-ll-***	***-l-***	***-ogue***	***-og***
organise	organize	counsellor	counselor	analogue	analog
realise	realize	travelling	traveling	dialogue	dialog

Others:

ageing vs. aging
mould vs. mold
aluminium vs. aluminum
sulphur vs. sulfur

I suspect you, the reader, have preferences from both British and American English. Writers do as well. It is increasingly common to see both British and American English in the same document. Provided the writer spells the same words consistently, they may choose either spelling according to their preference. However, writers should cater to the readers and must follow the style requirements of the instructor, employer, or publisher to whom they are submitting their work.

Exception: spell words that are part of a formal name according to the formal name, not the writer's preference.

A writer may prefer *center*, but would spell the "Centre of Excellence in Science Education" because that is its formal name.

Surprisingly, one of the major changes when preparing a "US Edition" or "International Edition" of a book involves changing the spelling and scientific units.

Latin phrases

Latin phrases are more common in older scholarly articles than they are today. They succinctly set the context of the information, but add a barrier to understanding if the reader does not know the Latin phrase. The use of Latin phrases is decreasing, which improves the readability of the work. If you chose to use Latin phrases, punctuate them as the English word.

Table 1.8 Latin phrases common in English prose. When abbreviations are given in parentheses, that abbreviation is used more commonly than the full phase.

Latin	English	Latin	English
ad hoc	to a specific task	non sequitur	it does not follow
bona fide	genuine; real	per capita	per person
ergo	therefore	per diem	per day
et alia (et al.)	and others (people)	per se	by itself; on its own
et cetera (etc.)	and others (examples)	post mortem	after death
id est (i.e.)	that is	prima facie	on the face of it
in situ	in the same place	sic	found in writing
in utero	in the womb	verbatim	identical; word-for-word
in vacuo	in a vacuum	vice versa	the positions being reversed
in vitro	in a test tube	vide infra	see below
in vivo	in a living organism	vide supra	see above
inter alia (i.a.)	among other things	videlicet (viz.)	that is to say; namely
a priori		from the former; from postulates	
ab initio		from the beginning; from first principles	
exempli gratia (e.g.)		for example	
ibidem (ibid.)		in the same place as before (in citations)	
loco citato (loc. cit.)		in the place cited (in citations)	
opere citato (op. cit.)		in the work cited (in citations)	
quod erat demonstrandum (Q.E.D.)		which was to be demonstrated/shown	

Conversely, legal procedures, medical illnesses, and biological taxonomy are founded in Latin. It is not likely that this will change so people studying these disciplines must learn sufficient Latin to understand and function in the discipline.

1.5 Common revisions

Recall the hallmarks of quality scientific prose: clarity, coherence, concision, and precision. As you write and review work, continuously ask yourself:

- What information does this word/phrase add?
- What other meaning can a reader infer?
- How can I rewrite this sentence to improve its readability?

Below are selected revisions you should consider making to improve the readability of your work. These are only representative revisions meant to prime your mind: as you write and review work, constantly monitor what you are doing to determine how best to convey the information.

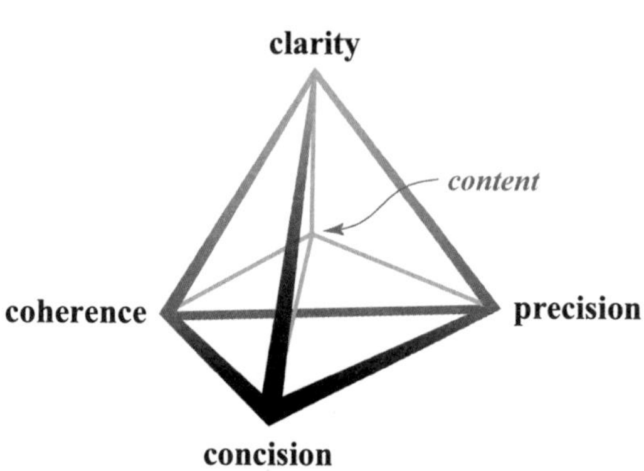

> *Omit needless words!* — William Strunk, Jr., *The Elements of Style*, 1918

> *Write with precision, clarity and economy. Every sentence should convey the exact truth as simply as possible.* — Instructions to Authors, *Ecology*, 1964

Jargon

Jargon is specialized technical language common to a scientific field, but not commonly used outside that field. The use of jargon introduces a barrier to understanding by non-specialist readers. The amount of jargon depends on the intended audience:

- Abstracts and documents for public audiences should not contain jargon.
- Documents for academic audiences (reports, essays, theses, oral and poster presentations) may contain jargon, but the jargon must be defined in terms understood by the intended audience.
- Documents for scientific audiences (journal articles) will contain jargon, but the jargon must be defined in terms understood by the intended audience.

Selecting appropriate words and phrases

The goal of scientific communication is to convey information as clearly and concisely as possible. Choose words and phrases so that the audience understands the work. Technical and scientific terms are appropriate for academic audiences, while common terms with similar meanings are more suitable for public audiences. Endeavoring to *wow* a public audience with highly technical language will have the opposite effect: the audience will not understand what you have written and may feel disdain for you and science. Table 1.9 lists scientific terms and near-equivalent common terms. See also the changes recommended to reduce tautology (page 54) and circumlocution (page 55).

If writing for a diverse audience, write for the simplest audience. Every reader will understand what you have written, and the academic reader interested in your work is usually comfortable contacting you to learn more about your work.

Table 1.9 Terms appropriate for scientific and public audiences.

Scientific	Public	Scientific	Public
accomplish	do	exhibit	show
accumulate	gather	fabricate	build
acquire	get	facilitate	help; simplify
activate	begin; start	hypothesis	educated guess
anticipate	expect	implement	begin; start
assist	help	initialize	start
commence	begin; start	initiate	begin; start
concept	topic	modify	change
conjecture	guess	objective	goal; aim
consider	think	optimum	best
construct	build	phenomenon	observation
currently	now	proactive	preventive
demonstrate	show	procure	get
discontinue	stop	represents	is
efficacious	effective	subsequently	after
elucidate	explain	sufficient	enough
endeavor	try	terminate	end; stop
evidence	fact; data	utilize	use

Some words have secondary meanings that may introduce ambiguity into the information you wish to convey. You want to use the correct scientific word to convey information. Below are some words that have specific meanings or multiple meanings. They should be used appropriately in your prose. The scientific terms on page 25 must also be used appropriately.

angle → perspective/viewpoint (defined term in mathematics)
concludes → shows/suggests/supports (research cannot prove something to be true)
maximize → increase/higher (are you sure it is a *maximum*?)
minimize → reduce/lower (are you sure it is a *minimum*?)
proves → shows/suggests/supports (research cannot prove something to be true)
sacrificed → killed (no ritual implication)
significant → <change> (defined term in statistics)
spectrum → <change> (defined term in spectroscopy)
unique → <change> (are you sure it is truly *unique*?)
weight, mass (both are defined terms; see page 70)

Additionally, some words have different connotations.

- Some words preferentially apply to one sex.

  beautiful vs. handsome

- Some words have legal implications.

  lacks ability vs. incompetent

- Some words have different degrees of formality.

  puking vs. vomiting vs. emesis

Contractions

A *contraction* is a shortened form of a word, substituting the missing letters and words with an apostrophe. Contractions are common in literary and informal prose, but should not be used in scientific prose. They add unnecessary complexity for readers whose first language is not English and can reduce the formality of your tone.

can't → cannot
dep't → department
don't → do not
he's → he is
I'll → I will
it's → it is
she's → she is
that's → that is; that has
there's → there is; there has
they're → they are
won't → will not
you're → you are

Commonly confused words

Affect or *effect*

Affect (verb)* means "to act on", "to influence", or "to produce a change in". Use *affect* if you can substitute *alter*, *modify*, *move*, or *transform* and the sentence still makes sense.

The wind affected the acceleration rate.

Effect (noun) means "the result of" or "the ability to produce results". Use *effect* if you can substitute *appearance*, *consequence*, or *outcome*, and the sentence still makes sense.

One effect of climate change is an increase in sea water levels.

Effect (verb)* means "to cause", "to bring about", "to accomplish", or "to make happen". Use *effect* if you can substitute *caused*, *implement*, *made*, or *make*.

We can effect ripening of bananas with ethylene.

Amount or *number* or *quantity*

Amount refers to something, usually singular, that is not quantified (water, sand, …) or not quantifiable (work, information, feelings, …).

A large amount of water is required for cooling.
A small amount of light leaked into the detector.

Number refers to something, usually plural, that can be quantified.

A large number of samples must be analyzed. (could also use *quantity*)
The virus infected a small number of computers.

Quantity refers to something, either singular or plural, that can be quantified.

A large quantity of samples must be analyzed. (could also use *number*)
The beaker contains a small quantity of waste. (could also use *amount*)

* *Affect* and *effect* as verbs:

Use *affect* when the object is the thing being changed.

The fire affected the temperature nearby. (Temperature is the object.)

Use *effect* when the object is the change itself.

The fire effected a change in the temperature. (Change is the object.)
He effected his escape with knotted bed sheets.
You should effect these changes on friday.

Among or *between*

Among refers to a collective or group containing many things.

A growing consensus among scientists is that sugar is as great a health risk as fat.
Tianna scored highest among her friends.

Between distinguishes, relates, or compares two or more things.

An agreement was reached between scientists to share data.

Because or *since*

Because introduces an explanation for an action.

The analysis will take three weeks because of the large quantity of data.

Since is associated with time.

I have been working on this since 2010.

Because of or *due to*

Because of is an adverb that modifies a verb.

Our field trials were cancelled because of weather.

Due to is an adjective that modifies a noun or pronoun; it is commonly preceded by the verb "to be". *Due to* is correct if you can substitute with *attributable to*, *caused by*, *resulting from*, or *supposed to*, and the sentence still makes sense.

The cancellation was due to weather.
The instruments are due to arrive today.

Can or *will*

Can means "is capable of".
Will means "is going to" or "is willing to".

Compared to or *compared with*

Compared to refers to similarities between objects.
Compared with refers to differences between objects.

Farther or *further*

Farther refers to physical distance.

The car is farther down the street.

Further refers to time or quantity.

How much further are we reading?

Few or *less*

Few refers to objects that can be counted (count nouns).

few students

Less refers to objects measured in bulk (mass nouns).

less rain

I or *Me* (see also *who* or *whom*)

Whether *I* or *Me* is used to identify the first person in a sentence depends on the context of the sentence. If everyone else is removed, which is the appropriate pronoun to use?

Will you help Jamie, Susan, and me clean the laboratory?
→ Will you help me clean the laboratory?
Jamie, Susan, and I are cleaning the laboratory.
→ I am cleaning the laboratory.

Formally: *I* is a subjective pronoun and *me* is an objective pronoun.

Imply or *infer*

Imply means "to state indirectly".

The preliminary report implies that damage was caused by defective material.

Infer means "to deduce".

I infer that we will be analyzing the construction material to identify the defective components.

If something is not stated directly, writers and speakers *imply* it with their prose, while the audience *infers* meaning from their words.

Its or *it's*

Its indicates possession.

The bird lost some of its feathers.

It's is the contraction of "it is".

It's a wonderful specimen.

Like or *such as*

Like denotes a comparison to one noun.
Such as denotes a comparison to more than one noun.

Loose or *lose*

Loose means "not tight" or "not attached".
Lose means "fail to retain" or "misplace".

Man or *male* *Woman* or *female*

Man/woman are nouns.

The man wearing the gray blazer.

Male/female are adjectives.

Female students.

Principal or *principle*

Principal (noun) means "the head or director", usually of a school; (adjective) means "highest importance" or "most important" in a group.
Principle (noun) means "a fundamental doctrine, tenet, or truth".

That or *which?*

That precedes a restrictive clause. Removing the clause changes the meaning of the sentence.

Test tubes <u>that have flared ends</u> cannot be used in the centrifuge.
Test tubes cannot be used in the centrifuge.

Which precedes a non-restrictive clause. Removing the clause removes information, but does not change the meaning of the sentence.

The samples were contaminated, <u>which resulted in erroneous results</u>.
The samples were contaminated.

Which is often preceded by a comma, whereas *that* is not.

Who or *whom* (see also *I* or *me*)

Who replaces "he" or "she" in a sentence.

Professor Leung is the person <u>who</u> you need to talk to.
<u>He</u> is the person you need to talk to.

Whom replaces "him" or "her" in a sentence.

To <u>whom</u> did you give the book?
I gave the book to <u>her</u>.

Formally: *who* is a subject pronoun and *whom* is an object pronoun.

Other commonly confused words

absorb/adsorb
accept/except
adapt/adopt
advice/advise
ascent/assent
assay/essay
bi____/semi____
censor/sensor
centered around/centered on
cite/sight/site
complement/compliment
continual/continuous
council/counsel
compose/comprise
device/devise
elicit/illicit
emit/omit
flammable/inflammable/nonflammable
generally/typically/usually
good/well
in/within
lay/lie
many/much
number of/total of
parameter/perimeter
shall/will
stationary/stationery
their/there/they're
to/too/two
translucent/transparent
vial/vile
your/you're

Gender neutrality

Two strategies to make text gender-neutral are:

- using a gender neutral term

  waiter/waitress → server
  manned → operated
  chairman/chairwoman → chair
  policeman → police officer

- changing the definition of a term

  actor/actress → actor
  guy/gal → guy
  he/she → they

The English language does not have a singular gender-neutral personal pronoun. "He/she" or "s/he" are used, but "they" and "their" are increasingly common as both a singular and plural pronoun.

A scientist must be objective in his/her analysis.
→ A scientist must be objective in their analysis.

Avoid misplaced modifiers

A modifier affects how another word is interpreted in a sentence. Modifiers must immediately precede or follow the word they modify. Misplacing the modifier introduces ambiguity into the sentence.

Who is eager to begin the experiment: the equipment or the students?

Eager to begin the experiment, the equipment was turned on by students.	(error)
Eager to begin the experiment, students turned on the equipment.	(correct) ✓

Who is going to the laboratory: they or the three classrooms?

They passed three classrooms on their way to the laboratory room.	(error)
On the way to the laboratory, they passed three classrooms.	(correct) ✓

Who/what is in pristine condition: the person or the samples?

Being in pristine condition after the trip, I believed the samples were not contaminated.	(error)
The samples were in pristine condition after the trip, so I believed they were not contaminated.	(correct) ✓

Who is in glass vials: the samples or the students?

The samples were provided to students in glass vials.	(error)
Students were provided samples in glass vials.	(correct) ✓

Did she say this on Friday? Or is she returning them on Friday?

The instructor said on Friday she would return our exams.	(error)
On Friday, the instructor said she would return our exams.	(correct) ✓
The instructor said she would return our exams on Friday.	(correct) ✓

Several words often cause misplaced modifiers: *almost*, *just*, *merely*, *nearly*, *only*, and *that*.

Did they present or not?

The research group nearly presented for 45 minutes.	(error)
The research group presented for nearly 45 minutes.	(correct) ✓

What is contaminated: the samples or the waste container?

Place the samples in the waste containers that are contaminated.	(error)
Place the samples that are contaminated in the waste containers.	(correct, passive) ✓
Place the contaminated samples in the waste containers.	(correct) ✓

How does the meaning change with the placement of the word "only"?

We only collect samples on Mondays.	(It is the only thing we do.)
We collect samples only on Mondays.	(Not on any other day.)
Only we collect samples on Mondays.	(No one else does.)

Avoid multiple adjectives and adverbs

Adjectives qualify nouns and pronouns, providing information such as "what kind", "which one", and "how many".

The automated sampler is very accurate.
The high-intensity light source saturated the sensitive detector.

Adverbs qualify verbs, adjectives, and all other components of a sentence, except nouns and pronouns. Adverbs provide information such as "how", "when", and "where".

The samples were analyzed immediately after collection.
Slowly add the acid into water while stirring vigorously.

Try to use only one adjective or adverb to qualify a phrase; more than this and it becomes difficult to determine what is being qualified. If more qualifiers are required, use a dash to separate the qualifiers from the phrase or separate the phrase into multiple sentences.

chemical healing suppression	(error)
suppression of healing by chemicals	(correct) ✓
chemical healing-suppression	(correct) ✓
Matrix ablated molecules were identified by spectroscopic analysis.	(error)
Spectroscopic analysis identified that the ablated matrix was molecular.	(correct) ✓

Avoid ambiguous pronoun references

A pronoun replaces a noun, but it must be clear what noun it replaces. Common ambiguous pronouns include *they*, *them*, *it*, *this*, *his*, and *her*.

Does "they" refer to the radio signals or airwaves?

Transmitting radio signals by satellite overcomes the problem of scarce airwaves and how they are used.	(error)
Transmitting radio signals by satellite overcomes the problem of scarce airwaves and how the airwaves are used.	(correct) ✓

Does "his" refer to the notes of the scientist or of the president?

The scientist gave the president his notes.	(error)
The scientist gave his notes to the president.	(correct) ✓

Which study does "it" refer to?

The second study was more in-depth than the first. It investigated the atmospheric lifetime of chlorinated compounds.	(error)
The second study was more in-depth than the first, which only investigated the atmospheric lifetime of chlorinated compounds.	(correct) ✓

Avoid anthropomorphism

Anthropomorphism is the attribution of living qualities to inanimate objects.* Anthropomorphism is not accepted in scientific prose because it imposes qualities onto objects that the objects do not have. In the extreme, readers may be left with the impression that an inanimate object is conscious of its surroundings, capable of making decisions, and capable of intentional actions.

The data suggests that LDL level and atherosclerosis are linked.	(anthropomorphic)
Analyzing the data, we discovered a correlation between LDL level and atherosclerosis.	(better) ✓
The storm ravaged the coast.	(anthropomorphic)
The intense wind, rain, and high waves damaged the coast.	(better) ✓
The fire licked the lower tree limbs.	(anthropomorphic)
The flame height was at the lower tree limbs	(better) ✓
My car refuses to start on cold mornings.	(anthropomorphic)
Cold weather prevents my car from starting.	(anthropomorphic)
My car does not start on cold mornings.	(better) ✓

Avoid double negatives

Readers, especially readers whose first language is not English, often misinterpret double negatives. Writers should remove them in their prose.

increasingly less common → decreasingly common *or* increasingly uncommon

Whenever possible, write positive relationships as positives.

It is not uncommon for … → It is common for …

It is not unreasonable to ask everyone to wear laboratory coats and safety glasses in the laboratory.	(poor)
It is reasonable to ask everyone to wear laboratory coats and safety glasses in the laboratory.	(better) ✓
The instrument not only measures temperature but also displays real-time data.	(poor)
The instrument measures temperature and displays real-time data.	(better) ✓

* Anthropomorphism originally meant the attribution of human qualities to animals and inanimate objects. Biological research continues to discover that animals exhibit many traditionally human qualities, including pleasure, empathy, and remorse.

The current scientific understanding is that organisms function with varying degrees of sentience, which affects the anthropomorphic traits that can be ascribed to them. For example, bacteria and viruses have minimal sentience. They do not choose to infect hosts, nor do they feel pleasure or remorse for their actions.

Avoid clichés and idioms

Clichés and idioms are expressions whose meaning is different from the words used and whose meaning requires the reader to have certain cultural knowledge. They confuse readers who try to interpret these expressions literally and add unnecessary words to prose.

A ***cliché*** is an overused expression that has lost its effectiveness. An ***idiom*** is an expression whose meaning is not consistent with the words in the expression. There are many idioms and clichés in the English language; they should not be used in scientific prose.

all other things being equal	(if the assumptions/hypotheses are correct)
at the end of the day	(in conclusion)
avoid ____ like the plague	(do not interact with ____)
break new ground	(start a new project)
bite your tongue; put a sock in it	(don't talk)
blinded by science	(confuse someone with technical information)
blow a fuse	(get angry)
burning the candle at both ends	(working too hard; burning out)
cry wolf	(false alarm)
different ball game	(an unrelated topic)
don't bite the hand that feeds you	(don't annoy your parents/boss)
get up on the wrong side of the bed	(in a bad mood)
go out on a limb	(make a speculative guess)
great minds think alike	(we both came up with the same great idea)
holy grail	(the highest possible goal)
in the black; in the red	(making a profit; losing money)
in the pipe	(a project actively being worked on; nearing completion)
it is raining cats and dogs	(it is raining very hard)
kick the bucket	(die)
last but not least	(another aspect or example)
leave no stone unturned	(very detailed investigation)
let the cat out of the bag	(accidentally reveal a secret)
light at the end of the tunnel	(make progress; things will get better)
light-years ahead	(leading others in the same area of research)
missing link	(a key piece of information or a key evolutionary species)

monkey business	(questionable actions)
not rocket science	(not very hard)
on the same wavelength	(thinking alike)
shed light on ____	(explain ____)
the tip of the iceberg	(a small portion of the problem is apparent)
think outside the box	(devise a creative solution)
throw ____ under the bus	(get someone else in trouble)
time is money	(time is valuable, don't waste it)
turn a blind eye	(ignore wrongdoing by others)
you've got your wires crossed	(you've misunderstood)
zero tolerance	(all violations, no matter how small, will be punished)

Avoid tautology

Tautology is the unnecessary repetition or needless clarification of a phrase.

1:30 AM in the morning → 1:30 AM
a specific example → an example
absolutely pure → pure
alternative choices → choices
an integral part → a part
are currently in the process of → are
brief summary → summary
collaborate together → collaborate
completely surrounded → surrounded
conclusive proof → proof
consensus of opinion → consensus
equal halves → halves
few in number → few
first began → began
first priority → priority
fully recognize → recognize
future plans → plans
grouped together → grouped
HIV virus → HIV
in equal halves → in halves
join together → join
mandatory requirement → requirement
mutual cooperation → cooperation
new innovation → innovation
oval in shape → oval
PAGE gel → PAGE
past experience → experience
perfectly clear → clear
PIN number → PIN
positive benefits → benefits
previously discovered → discovered
quite obvious → obvious
quite unique → unique
red in color → red
reverted back → reverted
related to each other → related
science of botany → botany
the data in fact shows → the data shows
the reason for this is because → because
true facts → facts
tuna fish → tuna
very unique → unique

Avoid circumlocution

Circumlocution is the use of excess words.

a decreased number of → fewer
a great deal of → much
a large number of → many
absolutely essential → essential
active consideration → consideration
afford an opportunity to → allow
alternative choice → choice
an increased length of time → longer
arrive at a decision → decide
at some time in the future → later
at the present time → now
based on the fact that → because
be cognizant of → know
be of assistance → assist
by means of → by/with
cause injuries to → injure
cognizant of → aware; know
come to the conclusion → conclude
consideration was given → I considered
conduct an analysis of → analyze
conducted a study of → studied
despite the fact that → although
due to the fact that → because
during the time that → while
few in number → fewer
final outcome → outcome
firmly commit → commit
first of all → first
for the purpose of → for, to
for the reason that → because, since
are found to be in agreement → agree
general consensus → consensus
grand total → total
has been shown to be → is
has the ability to → can
if conditions are such that → if
in accordance with → keeping to
in advance of → before
in all other cases → otherwise
in between → between
in conjunction with → with
in excess of → more than
in many cases → often
in order to → to
in other words → thus/hence/therefore
in regards to → about/regarding
in respect of → for
in the event that → if
in the neighborhood of → about
in the process of developing → developing
in view of the fact that → since, because
is based on → depends on
it is assumed that → if
it is often the case that → often
it is possible that → may
it may well be that → perhaps
it should be noted that → note that
it would appear that → apparently
longer time period → longer
made arrangements for → arranged
made the decision → decided
make a examination of → examine
mix together → mix
most unique → unique
of the same opinion → agreed
on a daily/regular basis → daily/regularly
on account of the fact that → as
original source → source
seal off → seal
separate entities → entities
small number of → few
smaller in size → smaller
spell out in detail → explain
take into consideration → consider
that have → with
the authors → I/we
the question as to whether → whether
the trend seems to suggest → perhaps
time period → time
two different methods → two methods
undertake a study of → study
until such time → until
unusual in nature → unusual
vast majority → most
very unique → unique
viable alternative → alternative
with a view to → to
without further delay → now

Avoid pleonasms

Pleonasms are words or phrases that add no information.

a case could be made
a total of
an interesting example is
are used to
as a matter of fact
at some future time
despite the fact that
has the effect of
human error
in the case of
in the form of
in this experiment/study
in the interests of
in the process of
interestingly
I should add that
it could be argued/said that
it goes without saying
it is agreed/clear/evident that
it is believed
it is important/interesting to note that
it is worth noting that
it may seem reasonable to suppose that
it must be noted/emphasized that
it should be mentioned that
it was found that
it would thus appear that
needless to say
nonetheless
so far, we have seen that
student error
the fact of the matter is that
the next thing to consider is
this has implications for
we wish to emphasize that
will have
with respect to
would like to

The word "that" can often be removed from sentences.

I see ~~that~~ you have started the …. → I see you have started the ….

Avoid vague and imprecise words

Scientific prose must be clear and concise. Vague and/or imprecise adjectives should be carefully reviewed to determine if a more specific phrase is appropriate. Some words sound knowledgeable, but are not specific. Often, they should be deleted and the sentence rewritten to be more precise.

DELETE

all, always
approximately
broad-based
clearly
comparatively, relatively
conceptualize
considerable
currently
definitely
dynamics
every
exciting, incredible, wonderful
exceptionally
extremely
fast, quick, slow
few, many, several, some
formalize
frequently, often
impact
infrastructure
interesting
large, small
like
literally
lots, little
multifaceted
never
obviously
parameters
poor, good, great
rather, relatively
recently
some
soon
quite
very

Consider the following examples.

The results were very significant because … (How much is "very")
A few samples showed signs of contamination. (How many? What signs?)

Additional resources …

… *on English grammar*

Definitions are adapted from http://www.dictionary.com

Fogarty M. Grammar Girl: Quick and Dirty Tips [internet]. Available from http://grammar.quickanddirtytips.com

Language and Learning Online [internet]. Available from http://www.monash.edu.au/lls/llonline/index.xml

Ross-Larson, B. Edit Yourself: A Manual for Everyone Who Works with Words. New York: W. W. North & Company; 1996.

Strunk W, White EB. The Elements of Style. 4th ed. Boston: Allyn & Bacon; 2000.

… *on communicating science*

Limerick P. Dancing with Professors: The Trouble with Academic Prose [essay; internet]. University of Colorado. Available from http://www.soc.umn.edu/~samaha/cases/limerick_dancing_with_professors.html

Sage L. Writing a clear and engaging paper for all astronomers, In: Astronomy Communication. Volume 290, chapter 13. Heck A, C Madsen, editors. Dordrecht (Netherlands): Kluwer. 2003.

Sheffield, N. Scientific Writing: Clarity, Conciseness, and Cohesion [internet]. Duke Scientific Writing Workshop. Available from https://cgi.duke.edu/web/sciwriting/resources/201108_DukeScientificWritingWorkshop.pdf.

Reconstructed paragraph: **3 1 5 4 2** (original on page 22)

Major earthquakes along a given fault line do not occur at random intervals because it takes time to accumulate the strain energy for the rupture. The time intervals are approximately constant because the rates of tectonic plates movement and strain accumulation along the fault are roughly uniform. *However, the time intervals will vary depending on the amount of strain required to start the earthquake and the amount of strain released during each earthquake.* Indeed, the time interval of major earthquakes along fault lines often varies by a factor of two. For example, the recurrence intervals along the southern segment of the San Andreas fault are 145 years with variations of several decades.

existing knowledge in topic position: \- - - - - - - - - -
new information in stress position: __________
topic sentence: <italicized>

You cannot become proficient at writing by only reading how to write. ***PRACTICE WRITING, PRACTICE REVIEWING,*** *and* ***PRACTICE EDITING*** *your work and others' work.*

You cannot become proficient at any activity — swimming, driving, communicating, research, … — by watching others. You must practice and, ideally, you must teach!

The best ways to uncover problems in your prose is to ***READ YOUR WORK ALOUD****. If you do not like the way it sounds, readers will not like the way it reads.*

Reading your work aloud will uncover problems with many aspects of your prose and help you achieve clarity, coherence, concision, and precision.

The first draft will never be perfect.
QUALITY PROSE REQUIRES REVIEW AND REVISION.

When preparing to write a document, budget time so that your work can be reviewed by yourself and others prior to submission.

Chapter 2. Communicating scientific information

There are established methods of communicating scientific information. By following these standards, your document will have increased clarity, coherence, concision, and precision, all of which improve readability and reduce the chance of misinterpretation.

Be aware that language is dynamic and constantly evolving. As you continue your education and career, you will observe changes in the grammar and style used to communicate information. In science, this evolution also includes new terms to explain new concepts.

2.1 Formatting common information

Capitalization

Common nouns denote a class or group with similar features, and are not capitalized. Instruments and chemicals are common nouns.

planet, country, city, apple, spectrometer, sodium chloride, …

Proper nouns denote a specific person, place, or thing; proper nouns are capitalized.

Jupiter, Russia, London, Macintosh, Ocean Optics spectrometer, …
out west vs. Western Canada

Some words can be both common and proper nouns, depending on the context.

university	(when referring to post-secondary education)
University	(when referring to a specific university)
the figures in the paper	(when referring to a collection of figures)
as shown in Figure 4	(when referring to a specific figure)
down the street	(when referring to an unnamed road)
down 5th Street	(when naming a specific road by name)

What constitutes a proper noun is changing as the language evolves. For example, the terms *internet* and *web* use to be proper nouns, but increased usage is converting them into common nouns. Days of the week are also transitioning to common nouns unless referring to a specific day.

Group meetings are every friday at 1600.
The conference begins Friday, 28 June 2013, at 1600.

Some words are proper nouns in some languages, but common nouns in others.

Monday (English) lundi (French)

Domain names are not case sensitive. However, the folder and file information in URLs is case sensitive.

www.nobelprize.org/ nobel_prizes/

not case sensitive case sensitive

www.nobelprize.org, www.NobelPrize.org, www.NOBELPRIZE.org (all valid)

Email addresses are not case sensitive.

alfred.nobel@nobelprize.org *or* Alfred.Nobel@NobelPrize.org

Titles and *headings* should only have the first letter and proper nouns capitalized.*

Date and time

Write the date in order of either increasing or decreasing duration. The year may be omitted if it is previously stated or obvious. If there is potential ambiguity, write the month in letters.

13 November 2012	13/11/2012	(increasing duration)
2012 November 13	2012/11/13	(decreasing duration)

We are meeting next Tuesday, 13 November.

Consider the date 03/06/09: this could be 03 June 2009, 09 June 2003 following the increasing/decreasing convention. If this convention is not followed, the date could also be in March or September. Ensure the dates you write are unambiguous.

Report time in the 24-hour clock to avoid ambiguity. The 24-hour clock uses leading zeros so that each of hours, minutes, and seconds each have two digits. A colon is used to separate hours:minutes:seconds, but may be omitted if reporting time as hours:minutes. If you use the 12-hour clock, use A.M. and P.M. to remove any ambiguity. A.M. and P.M. are written in small caps.

Classes start at 08:30. 21:35:08
Classes start at 0830. 9:35:08 P.M.
Classes start at 8:30 A.M.
The earthquake started at 1716 on 11 April 2011.

* Historically, writers capitalized every word in the title/heading to make it stand out from the surrounding text. This is no longer required because word processors have the ability to vary the font size and formatting to highlight headings.

Credentials and titles

Titles may be abbreviated with or without periods. Both forms are correct, but whatever style you choose, use it consistently. It is not correct to have one period at the end of a multiword abbreviation.

Dr *or* Dr.
PhD. → Ph.D. *or* PhD
MSc. → M.Sc. *or* MSc

It is not appropriate to duplicate a title.

Dr. Jill Pye, Ph.D. → Dr. Jill Pye *or* Jill Pye, Ph.D.

Addresses

Addresses are written on a minimum of three lines.

addressee	(first line)
<additional delivery information>	
street address	(second-last line)
community province/state postal/zip code	(last line)

Note that there is no punctuation in the last line: the community and province/state abbreviation are separated by one space, and the province/state abbreviation and postal/zip code are separated by four spaces. Formatting your address this way allows the post office's scanners to read the address; deviation from this standard can cause your mail to be misrouted.

Dr. Jordan Martin
Physics, University of Toronto
60 St. George Street
Toronto ON   M5S 1A7

To write an address in sentence form, separate each line with a comma.

Dr. Jordan Martin, Physics, University of Toronto, 60 St. George Street, Toronto ON   M5S 1A7, Canada

If shipping to another country, include the country on an additional line.

Mr. Sidney Gagnon
Science, Lynn High School
50 Goodridge Street
Lynn MA   01902
USA

Dr. Wolfgang Müller
Max Planck Institute for Physics
Föhringer Ring 6
80805 München
Germany

Phone numbers

Formatting of phone numbers varies around the world.

123-456-7890	North American
(123) 456-7890	North American
(123) 456-7890 x1234	North American, with extension
12 34 56 78 90	European

An increasingly common variant is to use periods to separate the blocks of numbers.

123.456.7890	North American
12.34.56.78.90	European

2.2 Formatting scientific information

The International Organization for Standards (ISO) publishes guidelines for communicating scientific information. ISO develops standards to promote good practices, open communication, and international trade. Each scientific discipline publishes a guide that focuses on the ISO standards common in its discipline. *Communicating Science* presents commonly used ISO standards. For details on specific disciplines, consult the following resources.

Biology: Quantities, Symbols, Units, and Abbreviations in the Life Sciences (IUMBM, IUBS)
Chemistry: Quantities, Units, and Symbols in Physical Chemistry (IUPAC)
Physics: Commission on Symbols, Units, Nomenclature, Atomic Masses and Fundamental Constants (IUPAP)
Engineering: <each engineering division — chemical, civil, electrical, mechanical — publishes their own style guide>

Mathematical equations

The formatting of the factors in mathematical equations provides information on the nature of the factor.*

- italicized serif font: *a*, *b*, *c*
    - *constants*: π (= 3.14159…), i ($=\sqrt{-1}$), e (electron charge)
    - *variables*: x, y, z, P, V, m, T
- upright serif font: a, b, c
    - *numbers*: 1, 2, 3, …
    - *functions*: cos, ln, e (as exponential function, e^{x}), Δ
    - *units*: mm, kg, MJ
    - *labels*: P_{V}, E_{A}, k_1, e (as electron), p (as proton), CO (carbon monoxide)
- bold italicized serif font: ***A***, ***B***, ***C***
    - *vectors* and *matrices*
- bold italicized sans-serif font: ***A***, ***B***, ***C***
    - *tensors*

* *Serif* fonts have stylized flairs at the end of the strokes that form the character. Times New Roman and Garamond are serif fonts: aA, bB, cC, …

Sans serif fonts do not have stylized flairs; the strokes that form the character end bluntly. Arial and Calibri are sans-serif fonts: aA, bB, cC, …

Examples

$$x = \frac{-b \pm \sqrt{b^2 - 4\,a\,c}}{2\,a}$$

$$\boldsymbol{A} = \begin{pmatrix} a_{11} & a_{12} \\ a_{21} & a_{22} \end{pmatrix}$$

$$c^2 = a^2 + b^2 - 2\,a\,b\cos(C)$$

$$d = d_0 + v\,t + \tfrac{1}{2}\,a\,t^2$$

$$F = m\,a$$

$$k(T) = A\,\mathrm{e}^{\frac{-E_A}{k\,T}}$$

$$\Delta_r G^\circ = -R\,T\ln(K)$$

$$P\,V = n\,R\,T$$

$$2\,\mathrm{H_2(g)} + \mathrm{O_2(g)} = 2\,\mathrm{H_2O(l)}$$

Numbers

When reporting quantities, use words for numbers up to and including ten, and numerals for numbers above ten.

We collected four samples. We collected 27 samples.

If a data set contains numbers below and above ten, use numerals for all numbers.

Samples 3, 7, and 12 showed signs of contamination.

When reporting data, use numerals. Numbers less than one require a leading zero and a decimal point. In all cases, the units should be treated as separate words and separated from the numeric value.

1.75 g 122.86 $
1.10 V 0.84 m/s

Exception: there is no space when reporting degrees, minutes, and seconds in angles and position

41° 43′ 55″ N, 49° 56′ 45″ W

Fractions less than one are written as either words or numerals. Fractions greater than one are written as numerals.

one-third *or* $^1/_3$ $^4/_3$ *or* $1\,^1/_3$

Numbers as labels are written as numerals.

Sample 3 Run 12

When writing two numbers consecutively, the first is written as a word and the second as a numeral, even if it violates the quantities rule.

This experiment requires twelve 250 mL volumetric flasks.

Uncertainty

The number of digits in a reported number indicates the precision to which that value is known. The values 4, 4.0, and 4.00 have different precision. In introductory courses, the uncertainty is often assumed to be ± 1 in the last reported digit.

4 ± 1
4.0 ± 0.1
4.00 ± 0.01

In advanced courses and in publications, the uncertainty in the number is explicitly reported. Uncertainty is estimated in measurements and carried through calculations using *propagation of uncertainty* (see Section 2.7).

4.00 ± 0.06

For large and small numbers, scientific notation simplifies the number and correctly indicates the uncertainty of the value.

65 000 000 years → $6.5 \cdot 10^7$ years → $(6.5 \pm 0.2) \cdot 10^7$ years
0.00784 mol/L → $7.84 \cdot 10^{-3}$ mol/L → $(7.84 \pm 0.05) \cdot 10^{-3}$ mol/L

Report values as decimals unless the fraction is absolute, then it is permissible to use fractions.

$^3/_{20}$ samples were contaminated.
15 % of the samples were contaminated.
Fifteen percent of the samples were contaminated.

Units

> If someone were to tell you that 5 = 11, you would question their intelligence. However, it is true that 5 kg = 11 lb. Units are necessary for understanding. All measurements must include both a numeral and units. Without both, there is no information in the data.

The metric system is the internationally accepted measurement system. There is less chance of confusion if all parties use the same units.

Some symbols are uppercase, but their names are lowercase.

mega, M	pressure, *P*
liter, L	temperature, *T*

Some scientific units are named after influential scientists. When writing the unit in full, write the name in lowercase. When writing about the scientist, capitalize the name.

kelvin, K	gauss, G
joule, J	becquerel, Bq

Units are not pluralized unless written in full.

1.0 kg, 7.5 kg	1.0 kilogram, 7.5 kilograms

Units are symbols, not abbreviations. There is no period after any metric unit.*

Numerals and units follow the rules of mathematics.

1.50 cm × 2.00 cm	*not* 1.50 × 2.00 cm
25 °C to 28 °C *or* (25 to 28) °C	*not* 25 to 28 °C
12.62 g ± 0.04 g *or* (12.62 ± 0.04) g	*not* 12.62 ± 0.04 g

To present complex units, use brackets or full mathematical notation.

J/(mol K) *or* J mol^{-1} K^{-1} *or* $\frac{\text{J}}{\text{mol K}}$ *not* J/mol K

Spacing within mathematical expressions

There is a space between every symbol in a mathematical formula: scalars, variables, functions, units, and operations (+, –, ×, ÷, =, …). In other words, punctuate each symbol as if it were a word.

$y = m\ x + b$	$x = 14\ \%$	$m = 14.65$ g NaCl
$b < 3$	$T = (310 \pm 2)$ K	$d = 0.274$ m

Exceptions:

$^{13}/_{20}$ $v = d/T$ kJ/mol	no space around the solidus (division symbol)
$^{1}/_{2}\,m\,v^2$ $^{12}_{6}$C e^{-x}	no space when the factor is super or subscripted
$H_2(g)$	no space between a label and the quantity it applies to
$F = m{\cdot}a$ $5.8{\cdot}10^{-4}$	no space around the dot used for multiplication
$T = (310 \pm 2)$ K	no space between the bracket and values inside the bracket
$t = 14.3$ °C	no space in the "degrees celsius" units†
$\theta = 53°\,31'\,12''$	no space when reporting degrees, minutes, and seconds

For numerals with four or more digits, use a *thin space* to separate the numeral into groups of three digits. Table 1.2 lists common types of spaces. Avoid using a comma, as it has a different meaning in Europe.

$F = 96\,485.336\,5$ C/mol	$h = 6.626\,069\,57{\cdot}10^{-34}$ J s

If you justify text, you will notice that your word processor changes the width of the space. To keep a normal width space in mathematical expressions, use a *non-breaking space* instead of a regular space.

* The imperial unit of inches is written as "in." to differentiate it from the word "in".

† "Celsius", as a unit, is still commonly spelled with a capital, despite the units convention presented on page 69.

Plural symbols and numerals

For symbols and abbreviations, add (’s).

The $\Delta_r H°$’s calculated for the system … CFC’s

For numerals, add (s).

1990s means the range from 1990 – 1999.
1990’s means something belonging to 1990.
’90s is an abbreviation of 1990s.

Mathematical expressions in sentences

Two practices are common when including mathematical expressions in sentences: using the unit symbols or writing them out in full. If the units contains more than two terms, the symbols must be used. When the units are written in full, a hyphen may be used to join the numeral and the units. Whichever practice you follow, be consistent throughout your work.

5.5 kg *or* 5.5 kilograms *or* 5.5-kilograms
a 25 mL aliquot *or* a 25-milliliter aliquot *not* a 25-mL aliquot
125 V/cm *or* 125 V cm^{-1} *or* 125 volts per centimeter
8.57 cm^3 *or* 8.57 cubic centimeters
1.26 mol/(L s) *or* 1.26 mol L^{-1} s^{-1} *not* 1.26 moles per liter per second

Variables and scientific terms can either be condensed or written in full:

The standard reaction enthalpy is the heat energy exchanged …
The heat energy exchanged, $\Delta_r H°$, …

The Fe^{3+} concentration is $2.83 \cdot 10^{-5}$ mol/L.
The iron(III) concentration is $2.83 \cdot 10^{-5}$ mol/L.

When used in sentences, punctuate all mathematical components according to their function in the sentence. Do not start a sentence with a numeral or variable.

We found that the activation energy of reaction 1 is (62.7 ± 0.8) kJ/mol.
We found that $E_{A,1} = (62.7 \pm 0.8)$ kJ/mol.

At 310 K, the enzymatic activity is ….
When $T = 310$ K, the enzymatic activity is ….

Mass versus weight

The terms *mass* and *weight* have different meanings in science than in society.

- *Mass* is measured in kilograms; the mass of an object is constant.
- *Weight* is a force and measured in newtons.

Weight and mass are related through the mathematical relationship

weight = *mass* × *acceleration*

Consider a mass of 75.0 kg

- standing at sea level: $g = 9.81$ m/s^2; $w = 7360$ N
- on thc top of Mount Everest: $g = 9.76$ m/s^2; $w = 7320$ N
- in a car accelerating horizontally at 5.00 m/s^2: the net acceleration is 11.01 m/s^2 and the weight is 8250 N
- in orbit around the earth: $g = 0.0$ m/s^2; $w = 0$ N

Even in scientific prose, *weight* is sometimes used incorrectly.

The weight of Sample 7 is 22.65 g.	(error)
The mass of Sample 7 is 22.65 g.	(correct) ✓
Weigh the samples before and after heating.	(error)
Measure the mass of the samples before and after heating.	(correct) ✓

Experimental labels

Experiments are conducted under varying conditions. Labels such as *condition A*, *condition B*, … contain no information about the experimental conditions and require the reader return to the experimental section to recall the experimental conditions. A better option is to assign meaningful labels to the experiments so that the reader is aware of the conditions throughout the paper.

An experiment that measures the effect of metals on plant growth rates:

Condition		
	A (low iron)	2.0 mmol/L $FeCl_3$ added nutrients
	B (high iron)	10.0 mmol/L $FeCl_3$
	C (low copper)	2.0 mmol/L $CuCl_2$
	D (high copper)	10.0 mmol/L $CuCl_2$
	⋮	

Original text: The plants exposed to condition C showed no statistical difference in growth compared to the control, whereas plants exposed to conditions A, B, and D were shorter and had less leaf area as tabulated in Table 7.

Improved text: The plants exposed to condition C (low copper) showed no statistical difference in growth compared to the control, whereas plants exposed to (low iron), (high iron), and (high copper) were shorter and had less leaf area as tabulated in Table 7.

2.3 Copyright and plagiarism

Copyright is a legal right granted to the creator[*] of a *work*, giving the creator exclusive rights to publish, produce, sell, and distribute the work. *Works* can be written (literary, academic), musical, dramatic (plays, concerts, movies, presentations), and artistic (paintings, photos, sculptures). Copyright law also allows the creator to transfer their copyright to others.

In the context of *Communicating Science*, the documents you prepare are academic works, and the presentations you give are dramatic works.[†]

In most countries, copyright exists the instant a work is produced. There is no requirement for any copyright symbol or statement, but it is common practice to include the copyright symbol, the creator, and the year the work was created: © Roy Jensen, 2014. The author of the work holds the copyright unless there is an agreement to the contrary or an employment relationship that transfers copyright to their employer.

The duration of copyright depends on the country. It could be as long as

- 95 years from the date created if the copyright holder is a company
- 95 years from the date of death if the copyright holder is a person

After the copyright has expired, the work becomes *public domain*, meaning that anyone can use the work for any purpose. However, you must still cite public domain works unless the information is *common knowledge*.

In general, you must ask permission from the copyright owner to use their work. However, there are certain situations when permission is not required. The *fair use* and *fair dealing* provisions (terminology varies by country) allow you to use sections of another creator's work without their permission, but you must still cite the work. While specific rights vary from country to country, fair use/dealing provisions apply when portions of a work are used for

- research and personal study
- non-profit instruction
- review and criticism
- news reporting
- satire and parody (limited countries)

[*] *Creator* applies to all works; *author* applies to written works.

[†] In law, there is no difference between a singer at a concert and you giving a presentation: both are dramatic works and covered by copyright.

An important consideration as to whether use qualifies as fair use/dealing is the amount of work copied and the reason for using it.

Copying an entire work completely for research, criticism, or parodying may be acceptable as fair use/dealing. For example, copying an entire poem for use in a book analyzing poetry may be acceptable. In addition, numerous comedy movies have scenes that parody other movies. Conversely, copying a small portion of a book and using it in another book is likely not fair use/dealing. For example, copying material out of a textbook so students do not have to purchase the textbook violates copyright laws.

> The research and personal study and news reporting provisions of fair use/dealing allow you to quote someone else in your work without their permission. However, you must still cite their work!

Plagiarism

Plagiarism is to knowingly pass off the original and distinctive ideas or work of another author as your own without crediting the author and when the context of such use expects the work to be cited.*

Consider the following text:

> A *hypothesis* is a proposed and testable explanation of phenomena. A good hypothesis explains the phenomena and predicts the outcome of future research. If the predictions are incorrect, the hypothesis is rejected. If the predictions are correct, the hypothesis gains strength and credibility. The predict–test–update hypothesis cycle continues to refine the hypothesis to more accurately explain the phenomena and to establish the applicable range of the hypothesis. Once understanding is achieved, the results are published. Publishing adds this information to the body of scientific knowledge for other scientists to evaluate, support, criticize, and build upon. If the hypothesis gains acceptance in the scientific community, it becomes a *theory*.

Plagiarism includes

- word-for-word (verbatim) copying without citing the source
- copying key words or phrases without citing the source

  > A *hypothesis* is a proposed and testable explanation of observed phenomena. A good hypothesis explains the phenomena and predicts the outcome of future research. If the predictions are incorrect, the hypothesis is rejected. If the predictions are correct, the hypothesis gains strength and credibility. If the hypothesis gains acceptance in the scientific community, it becomes a *theory*.

* Plagiarism is a complex, nebulous concept. The same material may require citation in one context, but not in another. For more information, see http://www.consol.ca/Plagiarism.pdf

- paraphrasing the idea without acknowledging its source

  A *hypothesis* is a proposed explanation of observed phenomena that predicts the outcome of future research. A hypothesis that correctly explains the phenomena and correctly predicts the outcome of future research is assumed to be correct and is published. If the hypothesis gains acceptance in the scientific community, it becomes a *theory*.

Context plays an important role in determining whether information should be cited. The established practice for persons preparing scholarly documents is to cite information taken from other sources. However, that same rigorous requirement does not apply in all contexts:

- Administrators writing policies review existing policies from other institutions and often copy sections verbatim or with minor modification, but rarely are these sources cited.
- In legal cases, judges prepare their decision using arguments written by attorneys, but without citing the attorney.
- Ghostwriting is common in publishing companies. Ghostwriters write novels that are published under the name of a popular author, increasing sales. Additionally, celebrities hire ghostwriters to write their memoirs and autobiographies.
- Academic instructors routinely create assignments and exams using questions taken from textbooks. The sources are rarely cited.
- Textbooks contain numerous real-world examples to illustrate the importance and applicability of the material in the text and questions. Only rarely are sources cited.

In these and many other contexts, the citation expectations differ from the rigorous expectations when preparing academic scholarly documents.

Common knowledge

Opposite plagiarism is the concept of ***common knowledge***. I did not develop the scientific method, nor did any living scientist or textbook author. Yet no current textbook cites a source when presenting the scientific method. The concept of the scientific method is *common knowledge* in the scientific community. Any scientist could prepare a reasonable statement on the scientific method without consulting a reference. However, if they choose to use or paraphrase other source(s), the source(s) must be cited. Similarly, the ideal gas equation, Green's theorem, Newton's laws of motion, and the Michaelis-Menten equation are common knowledge within the scientific community.

When does a concept go from requiring citation to becoming common knowledge? It depends on the person, their knowledge, the concept, and the work they are preparing. For example, the common knowledge of a scientist practicing in a discipline is different from a student learning the discipline. The common knowledge of a mathematician differs from that of a biologist. As a rule, if you look up information (other than scientific constants) and rely on that document to prepare your own document, that source must be cited. If in doubt, it is best to cite the source.

Copyright laws and the concept of plagiarism are changing as the world adapts to new technology and the information age. You should use the guidelines of your instructor, employer, or publisher when preparing scholarly works.

Non-copyrightable information

First, certain information is not copyrightable:

- ideas
- discoveries
- facts
- processes

In law, you may reproduce this information without citation. However, from an academic perspective, you should cite the source to give credit to the creator, to add credibility to the information you present, and to avoid allegations of plagiarism.

For example, consider a scholarly article that contains a procedure (a process) for setting up and conducting an experiment. Others can reproduce the procedure in their work, either verbatim, paraphrased, or with improvements. In law, the procedure does not require citation. In academia, the procedure requires citation. Conversely, you do not need to cite the source when looking up physical constants (facts).

Second, common symbols and layouts are not copyrightable.

- *symbols*: WHIMS and scientific symbols, stop and yield signs, washroom signs, …
- *layouts*: periodic table, calendar, …

Third, an image of a public domain work is also in the public domain. For example, you can use a picture of the Mona Lisa or the Magna Carta without citing the photographer who took the photo.

Public copyright licenses

The advent of the internet and the plethora of documents available online has revealed a limitation of existing copyright laws: it may be difficult to identify and obtain permission from the copyright holder.

To address this limitation using the existing laws, public copyright licenses were proposed that allow the copyright holder to grant others permission to use the work with certain conditions. One type of public copyright license is the Creative Commons license. The Creative Commons organization proposed four terms that can be selected by the copyright holder and forms the foundation of the Creative Commons, 🅭, license. Table 2.1 summarizes these terms. All licenses require the *attribution* term plus any combination of the other terms. With the exception of work licensed with the *no derivative* condition, you may modify and adapt the original work into your own (called a ***derivative work*** in law).

Table 2.1 Possible terms that a creator could require of licensees using a Creative Commons license.

Licence terms	
attribution (BY)	Licensees must give appropriate credit to the creator, provide a link to the CC license, and indicate if changes were made. (required)
share alike (SA)	Licensees must distribute the derivative work under the same licence as the original work. (optional)
non-commercial (NC)	Licensees may not use the creator's work for commercial purposes. (optional)
no derivatives (ND)	Licensees may only use the creator's work verbatim and in its entirety. (optional)

Source: Creative Commons, www.creativecommons.org

Not all Creative Commons licenses are compatible. It is your responsibility to ensure you use the Creative Commons work of multiple other authors legally and that it is compatible with how you plan to license your work.

Another option a creator can use to make their work useable by others is to put their work into the public domain, which allows anyone to use the work for any purpose.

2.4 Citations

A citation provides a reader with sufficient information to locate a work used by the author. This ensures that people receive credit for their knowledge and allows the reader to obtain more information about topics they are interested in. By knowing the source, a reader can also assess the credibility and potential biases of the information. For example, Greenpeace and OPEC (a consortium of petroleum producers) may have significantly different perspectives on climate change and pollution.

Citation styles

Numerous organizations have developed style guides. Some common style guides include*

- Modern Language Association (MLA) style, commonly used in literature, arts, and humanities.
- Chicago style, commonly used in non-scholarly books, magazines, and newspapers. (A variant of Chicago is the Turabian style.)
- American Medical Association (AMA) style, commonly used in medical and health disciplines.

Several scientific organizations have developed style guides as well.†

- The Council of Science Editors (CSE) style‡
- American Chemical Society (ACS) style
- American Institute of Physics (AIP) style
- Institute of Electrical and Electronics Engineers (IEEE) style
- American Psychological Association (APA) style§

* Additional information and resources on
MLA: http://www.MLA.org/style/
Chicago: http://www.Chicagomanualofstyle.org
AMA: http://www.AMAmanualofstyle.com

† Additional information and resources on
CSE style: http://www.CouncilScienceEditors.org
ACS: http://chemistry.library.wisc.edu/writing/acs-style-guidelines.html
AIP: http://www.AIP.org/pubservs/style/4thed/toc.html
IEEE: http://www.IEEE.org (search for 'style guide'; there are several)
APA: http://apastyle.APA.org

‡ The Council of Biology Editors (CBE) became the Council of Science Editors (CSE) in 2000 to reflect the growing diversity of their membership.

§ APA style is commonly used in anthropology, economics, education, political science, psychology, sociology, and other social sciences.

No one style is better than any other. Consequently, there is little interest in switching to a common style. An instructor, employer, or publisher will select a style so that information is communicated in a consistent manner. As a writer, you must follow their style requirements when submitting documents to them.

Style guides address all aspects of document formatting, from text to tables to citations. However, there is significant commonality and flexibility among formatting guidelines, so the focus of instructors is often on formatting citations.

The citation format in *Communicating Science* is based on the Council of Science Editors (CSE) style manual.* The text box on page 80 and the Variations section on page 86 provide information on and rationalize alternate citation styles.

Citation formats

Citations are commonly ordered *alphabetically* or *numerically*, which introduces two common citation formats.

The ***alphabetical*** (***name-year***; ***Harvard***) format uses parenthetical references in the text body and orders the references alphabetically by author at the end of the document, in a section commonly called *Bibliography* or *Works cited.*

(Author Year) *or* (Author Year, page)	(in text)
<full citation to original work>	(in *Bibliography*)

The ***numerical*** (***citation-sequence***; ***Vancouver***) format uses numbers in the text body and orders the references numerically at the end of the document in a section commonly called *References*.

(#) *or* [#] *or* $^{\#}$	(in text)
#. <full citation to original work>	(in *References*)

* Adapted from

Council of Science Editors. Scientific Style and Format: The CSE Manual for Authors, Editors, and Publishers. 7th ed. Reston, VA: Rockefeller University Press; 2006.

For resource types not in the CSE Manual of Style, the principles therein were applied to create a consistent citation format.

Details on the *alphabetical* format

This format is commonly used in biology and environmental science.

There are several nuances of the in-text component of citations:

- If there is one author, list the author's last name and the year of publication.

  (Jones 2004)

- If there are two to three authors, list them together.

  (Jones, Nguyen, and Robinson 2008)

- If there are more than three authors, list only the first author and add 'et al.' to indicate the existence of additional authors.

  (Martinez et al. 2009)

- If required to differentiate between publications, more than one author can be listed.

  (Martinez, Carter, et al. 2011)
  (Martinez, MacDonald, et al. 2011)

- If multiple authors have the same last name and publish in the same year, include sufficient initials to differentiate the authors.

  (Smith L 2007)
  (Smith T 2007)

- If an author has published more than once in a given year, use a, b, c, … after the year to sequence the publications chronologically.

  (Torres 2006a)
  (Torres 2006b)

- If there are multiple citations, list them chronologically and separated by semicolons.

  (Müller 1997; Bateman 2000; Smith T 2007)

- If there is no date on the publication, state this in the citation.

  (Chen [no date])

- The page number is included when citing a book or when using a direct quotation from another resource.

  (Godbersen 2004, p. 265)
  (Shakur 2009, pp. 183–186)

If the author's name can be incorporated into the text, then only the year needs to be in parentheses.

Martinez et al. (2009) concluded that …

Details on the *numerical* format

This format is commonly used in chemistry, physics, and mathematics.

In the text, the reference number is either superscripted[#] or enclosed in brackets: (#) or [#]. There are several nuances of the in-text component of citations:

- Place the number close to the information being referenced or to the author of the information.

  Westwood[1] explored the effect of parasitic weeds on native …
  Parasitic weeds decreased the growth rate and lifespan of native vegetation.[1]

- Number citations in order of first occurrence.
- If there are multiple citations to the same information, number them chronologically.
- Separate multiple in-text citations using commas, but without spaces.
- Join more than two consecutive citations with an en-dash.

  The mass spectrometer is a powerful instrument that has applications in chemical analysis,[1–4] isotopic analysis,[5,6] dissociation dynamics,[2,7] and proteomics.[3,5,10–12]

  … dynamical status of galaxy groups and clusters [3,4]. The most active supercluster nuclei contain more early-type galaxies [5–8].

Full citations

> The differences between citation formats are cosmetic. There is no universally accepted citation format and all citations convey similar information. However, you must use the citation format dictated by the instructor, employer, or publisher to whom you are submitting your work.

The format for end-of-document citations is nearly the same in both the alphabetical and numeric styles. The only difference is the location of the publication year.

In the citation, the article title must be capitalized and spelled the same as the source, even if it violates the guidelines you are following for capitalizing titles and headings in your document.

Punctuation in citations serves specific purposes.

- Periods indicate the end of a component and the end of the citation. Periods are not used after the initials nor at the end of the citation if it ends with a URL or DOI (see page 86).
- Commas are used within a component to separate multiple authors, separate the author's name and role, and separate levels within an

organization. Commas are not used to separate the last name and initials.

- Semicolons separate related items within a component, such as the publisher and publication year.
- Colons separate the title from the subordinate title and the publisher and from the place of publication.

Citations must contain sufficient information to locate the work, but often contain more information than required. This redundancy allows a resource to be found even if there is an error in the citation. In addition, not all the information may be known for a citation, so that information may be omitted without losing the ability to locate the work.

> Numerous factors affect the citation format adopted by an organization. For journal publishers, space is a big factor, especially when every article cites tens of references. In developing a citation style, publishers endeavor to reduce the space required for citations by omitting information. The following three citations all provide sufficient information to find the reference.
>
> Ford AR, Burns WA, Reeve SW. Rotational Analysis of FTIR Spectra from Cigarette Smoke. Journal of Chemical Education. 2011;81(6):865–867.
>
> Ford AR, Burns WA, Reeve SW. J. Chem. Ed. 2011;81(6):865.
>
> J. Chem. Ed. 2011;81:865.
>
> While the first two have some redundancy, the last does not. An error in the last citation would make it impossible to find the article.

Book with author(s)

Author(s). Year.[alpha] Title. Edition. Place of publication: Publisher; Year.[numeric]

alpha: Voet D, Voet J. 2010. Biochemistry. 4th ed. New York: Wiley.

numeric: #. Voet D, Voet J. Biochemistry. 4th ed. New York: Wiley; 2010.

Book with editor(s)

alpha: McCuen G, editor. 1987. Our endangered atmosphere: global warming and the ozone layer. Hudson (WI): Gary E. McCuen Publications.

numeric: #. McCuen G, editor. Our endangered atmosphere: global warming and the ozone layer. Hudson (WI): Gary E. McCuen Publications; 1987.

Electronic book

Author(s). Year.[alpha] Title [internet]. Edition. Place of publication: Publisher; Year[numeric] [cited <date>]. Available from: <URL>

(The author is an organization in this example.)

alpha: Board on Mathematical Sciences and Their Applications (BMSA), Engineering and Physical Sciences (DEPS). 2012. Fueling Innovation and Discovery: The Mathematical Sciences in the 21st Century [internet]. Washington: National Academies Press [cited 04 July 2012]. Available from: https://download.nap.edu/catalog.php?record_id=13373

numeric: #. Board on Mathematical Sciences and Their Applications (BMSA), Engineering and Physical Sciences (DEPS). Fueling Innovation and Discovery: The Mathematical Sciences in the 21st Century [internet]. Washington: National Academies Press; 2012 [cited 04 July 2012]. Available from: https://download.nap.edu/catalog.php?record_id=13373

Book chapter, for books where each chapter has a different author

Author(s). Year.[alpha] Chapter title, in: Book title. Edition. Chapter <#>. Editor(s), editor(s). Place of publication: Publisher; Year.[numeric]

alpha: Cook HJ. 1996. Physicians and natural history, in: Cultures of natural history. Chapter 6. Jardine N, Spary EC, Secord JA, editors. New York: Cambridge.

numeric: #. Cook HJ. Physicians and natural history, in: Cultures of natural history. Chapter 6. Jardine N, Spary EC, Secord JA, editors. New York: Cambridge; 1996.

Journal article

Author(s). Year.[alpha] Article title. Journal title. Year;[numeric]Volume(Issue):Page range.

alpha: Chen Z, Ju Liu Z, Sheng Y. 2011. Dirac structures of omni-lie algebroids. International Journal of Mathematics. 22(8):1163–1185.

numeric: #. Chen Z, Ju Liu Z, Sheng Y. Dirac structures of omni-lie algebroids. International Journal of Mathematics. 2011;22(8):1163–1185.

Online journal article

Author(s). Year.[alpha] Article title. Journal title [internet]. Year[numeric] [cited <date>];Volume(Issue):Page range. Available from: <URL>

alpha: Halford NG, Curtis TY, Muttucumaru N, Postles J, Mottram DS. 2011. Sugars in crop plants. Annals of Applied Biology [internet] [cited 14 November 2011];158(1):1–25. Available from: http://www.blackwellpublishing.com/journal.asp?ref=0003-4746

numeric: #. Halford NG, Curtis TY, Muttucumaru N, Postles J, Mottram DS. Sugars in crop plants. Annals of Applied Biology [internet] 2011 [cited 14 November 2011];158(1):1–25. Available from: http://www.blackwellpublishing.com/journal.asp?ref=0003-4746

Conference proceedings article

Author(s). Date.[alpha] Article title. In: Editor(s). Proceedings title; Place of conference. Place of publication: Publisher; Date.[numeric] Page range.

alpha: Lee DJ, Bates D, Dromey C, Xu X, Antani S. 26–27 June 2003. An imaging system correlating lip shapes with tongue contact patterns for speech pathology research. In: Krol M, Mitra S, Lee DJ, editors. Proceedings of the 16th IEEE Symposium on Computer-Based Medical Systems; New York. Los Alamitos (CA): IEEE Computer Society. 307–313.

numeric: #. Lee DJ, Bates D, Dromey C, Xu X, Antani S. An imaging system correlating lip shapes with tongue contact patterns for speech pathology research. In: Krol M, Mitra S, Lee DJ, editors. Proceedings of the 16th IEEE Symposium on Computer-Based Medical Systems; New York. Los Alamitos (CA): IEEE Computer Society; 26–27 June 2003. 307–313.

Thesis, dissertation, or essay

Author. Year.[alpha] Title [designation]. Place of publication: Publisher; Year.[numeric]

alpha: Dettmers JM. 1995. Assessing the trophic cascade in reservoirs: the role of an introduced predator [Ph.D. dissertation]. Columbus (OH): Ohio State University.

numeric: #. Dettmers JM. Assessing the trophic cascade in reservoirs: the role of an introduced predator [Ph.D. dissertation]. Columbus (OH): Ohio State University; 1995.

Technical report

Author(s). Year.[alpha] Title. Report number. Place of publication: Publisher; Year.[numeric]

alpha: Moray NP, Huey M. 1988. Human factors research and nuclear safety. Report #NRC-04-86-381. Washington: National Academies Press.

numeric: #. Moray NP, Huey M. Human factors research and nuclear safety. Report #NRC-04-86-381. Washington: National Academies Press; 1988.

Course material

Author(s). Year.[alpha] Course. Title. Place of publication: Publisher; Year.[numeric]

alpha: Chen W. 2008. Math 133. First year calculus. Sydney (AU): Maquarie University.

numeric: #. Chen W. Math 133. First year calculus. Sydney (AU): Maquarie University; 2008.

Magazine

Author(s). Date.[alpha] Article title. Magazine title. Date;[numeric]Volume(Issue):Page range.

alpha: Losos JB. March 2001. Evolution: A lizard's tale. Scientific American. 284(3):64–69.

numeric: #. Losos JB. Evolution: A lizard's tale. Scientific American. March 2011; 284(3):64–69.

Newspaper article

Author(s). Date.[alpha] Article title. Newspaper title. Date;[numeric]Section:Page (column).

alpha: Chang K. 31 June 2000. Two stars collide: a new star is born. New York Times. Sect. F:1 (col. 4).

numeric: #. Chang K. Two stars collide: a new star is born. New York Times. 31 June 2000;Sect. F:1 (col. 4).

Encyclopedia

Author(s). Year.[alpha] Article title. In: Encyclopedia title. Volume. Place of publication: Publisher; Year.[numeric] Page range.

alpha: Robertson MP. 2009. Origins of life: emergence of the RNA world. In: Wiley encyclopedia of chemical biology. Volume 3. Hoboken (NJ): Wiley. 517–528.

numeric: #. Robertson MP. Origins of life: emergence of the RNA world. In: Wiley encyclopedia of chemical biology. Volume 3. Hoboken (NJ): Wiley. 2009; 517–528.

Page or document from a website

Author(s). Year.[alpha] Page title. Website title [internet]. Place of publication: Publisher; Year[numeric] [cited <date>]. Available from: <URL>

alpha: Lai A. 2010. Somnambulism (sleepwalking): asleep with your eyes wide open. End your sleep deprivation [internet]. Stanford (CA): Stanford Sleep and Dreams; [cited 03 July 2011]. Available from: http://www.end-your-sleep-deprivation.com/somnambulism.html

numeric: #. Lai A. Somnambulism (sleepwalking): asleep with your eyes wide open. End your sleep deprivation [internet]. Stanford (CA): Stanford Sleep and Dreams; 2010 [cited 03 July 2011]. Available from: http://www.end-your-sleep-deprivation.com/somnambulism.html

Presentation

A presentation may only be cited if you have a copy of the presentation to confirm what was presented.

Author(s). Date.[alpha] Presentation title. Conference; Place of conference. Date.[numeric]

alpha: Lofgreen JE, Browning CS, Dicks AP. 26–30 Mary 2012. Teaching Scientific Writing in a First-Year Chemistry Laboratory. Canadian Society for Chemistry conference; Calgary (AB).

numeric: #. Lofgreen JE, Browning CS, Dicks AP. Teaching Scientific Writing in a First-Year Chemistry Laboratory. Canadian Society for Chemistry conference; Calgary (AB). 26–30 Mary 2012.

Personal communication (meeting, interview, email, memo)

There must be documents to support any personal communication. Documentation could be a copy of the information that was discussed or an email summarizing the discussion.

Author. Date.[alpha] Personal communication. Affiliation. Date.[numeric]

alpha: Tarver, CM. 14 August 2004. Personal communication. Lawrence Livermore National Laboratory.

numeric: #. Tarver, CM. Personal communication. Lawrence Livermore National Laboratory. 14 August 2004.

Citing multimedia

Multimedia is everything other than text in your work: tables, figures, images, graphs, audio, and video.

You must cite multimedia created by another author, but the citation is usually at the location of the multimedia. For tables, figures, images, and graphs, the citation is commonly in the caption or in a footnote on the same page. For audio and video, the citation is commonly in a footnote on the same slide as the audio or video.

If you are using multimedia under a fair use/dealing exception, add the following component to the normal citation.

- If the multimedia is identical to the original, write the citation as,

  Source: <full citation>

- If the multimedia is modified from the original to better suit your work, write the citation as,

  Adapted from <full citation>

  Adapted from Diderot T. Arborescence – Physiologie [internet]. 2008 [cited 05 February 2012]. Available from http://picasaweb.google.com/taa.diderot/ArborescencePhysiologie

- If you are presenting data in another format — for example, taking data from a source and plotting it in your work — write the citation as,

  Source: data from <full citation>

  Source: data from Statens energimyndighet. 2010 Energy in Sweden - facts and figures 2010. Table 48.

If you have obtained permission to use another creator's multimedia, the copyright owner has the right to dictate the location and format of the citation. (For example, see the images on page 8, 116, 205, and 220.) If they do not specify a citation format, cite the multimedia as above. Add the following phrase to the end of the citation:

Used with permission.

Additional information added to citations

Online resources

References to online resources add three components to the citation:

- the phrase "[internet]" after the title
- the date cited
- the URL at the end of the citation. (There is no period after the URL.)

Additionally, keep a copy of the resource so that you can easily refer to it or provide it to others if required. If the resource is a website, print the relevant page(s) as a PDF.

Articles found in a database

Electronic databases index published work. If you find a reference using a database, acknowledge the database in the citation by adding the following phrase to the end of the citation:

Retrieved on <date> from <database name>.

Digital object identifiers

Scientific journals may assign a digital object identifier (DOI) to online articles. The DOI is a short unique identifier that links to the article through the DOI database. If DOIs are available, use them instead of the URL at the end of the citation.

Variations

As mentioned previously, citation formats vary tremendously, but all provide sufficient information for another person to locate the reference. As an author, you must use the citation style required by your instructor, employer, or publisher.

Some citation formats require the author's first name in the citation.

Voet, Donald; Voet, Judith. 2010. Biochemistry. 4th ed. New York: Wiley.

Other formats have different internal punctuation and apply italic and bold formatting to title, year, volume, and issue. Two examples are

American Institute for Physics (AIP) style

M. Tsujimoto, H. Minami, K. Delfanazari, M. Sawamura, R. Nakayama, T. Kitamura, T. Yamamoto, T. Kashiwagi, T. Hattori, and K. Kadowaki. Terahertz imaging system using high-T_c superconducting oscillation devices *Journal of Applied Physics* **111**, 123111 (2012).

American Chemical Society (ACS) style

Son, H.J.; Wang, X.; Prasittichai, C.; Jeong, N.C.; Aaltonen, T.; Gordon, R. G.; Hupp, J.T. Glass-Encapsulated Light Harvesters: More Efficient Dye-Sensitized Solar Cells by Deposition of Self-Aligned, Conformal, and Self-Limited Silica Layers. *Journal of the American Chemical Society*. **2011**, *134*, 9537.

Using material without permission

The same text and multimedia are often found on multiple websites. It is sometimes challenging to determine the source of the material and challenging to obtain permission from the copyright holder to use their material. If the author cannot be determined, state in a footnote or in the caption that the information is available from multiple sources online and there is no clear indication of whom the original author is.

[Page 299] An internet search for "science rubric" returns a plethora of rubrics. Many of them are duplicates of each other and there is no indication of the original author. The rubrics in this section were adapted from those found online, but it is not practically possible to identify and credit the original authors.

If you do identify the copyright holder, but they do not respond, you should assume that they do not grant permission and you should not use their material.

2.5 Presenting scientific data: tables

Tables are a natural way to record and present data in a convenient and logical format. Trends are more-readily observable when tabulated. Electronic tables can also be manipulated: columns added and deleted, calculations done, and the table formatted for publication.

Both word processors and spreadsheets are capable of making publication-quality tables. Examples of quality tables are found in *Communicating Science*, textbooks, and in published scholarly articles.

Scientific tables have

- clear and concise headings
- numerical data formatted to clearly and accurately present the data, with appropriate significant digits
- units transferred to the headings
- a minimum number of horizontal and vertical dividing lines

Tables are numbered sequentially and should be referred to in the text. If possible, integrate the reference into the sentence.

The results are tabulated in Table #.
The data in Table # summarizes …

Tables are captioned at the top. The caption concisely explains the data in the table and provides brief details on any information in the table that may be difficult to understand. The caption should be a maximum of three sentences. A table and caption should contain sufficient information to be understandable without the reader having to reference the text.

A recurring example used in sections 2.5 and 2.6 is an iron analysis based on the reaction of iron(III), Fe^{3+}, and thiocyanate, SCN^-, to form iron(III) thiocyanate, $Fe(SCN)^{2+}$.

$$Fe^{3+}(aq) + SCN^-(aq, excess) \rightleftharpoons \underbrace{Fe(SCN)^{2+}}_{\text{red}}(aq)$$

Absorbance spectroscopy measures the red intensity, which correlates to the iron(III) thiocyanate concentration in the original sample. A calibration curve is created using standards with known iron(III) concentrations.

Table 2.2 presents the raw data collected from the calibration standards and an unknown sample.

Table 2.2 Absorbance at 450 nm of iron(III) standards and unknown. Iron(III) exists as iron(III) thiocyanate.

label	$\mathbf{Fe^{3+}}$ **/(mol/L)**	A **450 nm**
	standards	
std. 1	$6.73\cdot10^{-5}$	0.593
std. 2	$5.07\cdot10^{-5}$	0.475
std. 3	$3.39\cdot10^{-5}$	0.336
std. 4	$1.68\cdot10^{-5}$	0.199
	unknown	
unk. A	$3.25\underline{0}\cdot10^{-5}$	0.325
	$3.35\underline{2}\cdot10^{-5}$	0.333
	$3.28\underline{8}\cdot10^{-5}$	0.328

When tabulating data, transfer the units to the heading following the rules of mathematics. In Table 2.2, the iron(III) concentration has units of mol/L. Absorbance is dimensionless.

$$Fe^{3+} = 6.73\cdot10^{-5}\,\frac{\text{mol}}{\text{L}} \quad\Rightarrow\quad \frac{Fe^{3+}}{\text{mol}/\text{L}} = 6.73\cdot10^{-5} \tag{2.1}$$

This can also be written on a single line with brackets used to remove the ambiguity in the units.

$$\underbrace{Fe^{3+}\ /(\text{mol}/\text{L})}_{\text{table heading}} = \underbrace{6.73\cdot10^{-5}}_{\text{table entry}} \tag{2.2}$$

Additionally, the order of magnitude can also be moved to the heading:

$$Fe^{3+} = 6.73\cdot10^{-5}\,\frac{\text{mol}}{\text{L}} \quad\Rightarrow\quad \frac{Fe^{3+}}{10^{-5}\ \text{mol}/\text{L}} = 6.73$$

$$\underbrace{Fe^{3+}\ /(10^{-5}\ \text{mol}/\text{L})}_{\text{table heading}} = \underbrace{6.73}_{\text{table entry}} \tag{2.3}$$

The replicate measurements of the unknown provide a mechanism for estimating the accuracy of the data. Two common calculations are the average and standard deviation. (Section 2.7 provides more information on the statistical analysis of data.)

Unknown A

$$\left.\begin{aligned} \textit{average}: & \ 3.30\cdot 10^{-5}\ \frac{\text{mol}}{\text{L}} \\ \textit{standard deviation}: & \ 5\cdot 10^{-7}\ \frac{\text{mol}}{\text{L}} \end{aligned}\right\} \ (3.30\pm 0.05)\cdot 10^{-5}\ \frac{\text{mol}}{\text{L}}$$

Table 2.3 presents the calibration and unknown data that would be in the final report. The raw data (Table 2.2) may be in an appendix to the report.

Table 2.3 Absorbance at 450 nm of iron(III) standards and unknown. Iron(III) exists as iron(III) thiocyanate. The uncertainty is the standard deviation.

label	$\mathbf{Fe^{3+}}$ **/(mol/L)**	$A_{450\ \text{nm}}$
	standards	
std. 1	$6.73\cdot 10^{-5}$	0.593
std. 2	$5.07\cdot 10^{-5}$	0.475
std. 3	$3.39\cdot 10^{-5}$	0.336
std. 4	$1.68\cdot 10^{-5}$	0.199
	unknown	
average:	$(3.30 \pm 0.05)\cdot 10^{-5}$	

2.6 Presenting scientific data: figures, images, and graphs

Presenting information as a figure, image, or graph provides an overview of the information and allows a reader to visualize trends in data. A glancc at a well-prepared figure, image, or graph may interest a reader to read the entire work.

figure: (schematic, diagram, line drawing) a pictorial representation of an object or process

image: (photograph, realistic drawing) a visual reproduction

graph: (chart) a representation of the quantitative relationship between two or more quantities, or a mathematical relationship

While there is a formal difference between figures, images, and graphs, many people commonly use *figure* to refer to all three, and *figure* is used in captions of figures, images, and graphs.

Figures provide a visual presentation of complex information, which aids in understanding. Figures are commonly used to present information about the experimental method, experimental setup, and the data and analysis. Figures should only contain details relevant to the project. A challenge in preparing figures is determining what to include and how to prepare the figure to best convey the desired information. In general, simpler figures are better than complex figures.

If you are using a figure produced by someone else, ensure that you have permission to use it or that you are using it under one of the fair use/dealing provisions (see Section 2.3).

Figures should be between half and a full page in size. When making figures and graphs, use a vector format so that it scales to any size. (Vector formats also have a smaller file size.) Images, such as photographs, are raster and limited by their resolution. Ensure the printed image resolution is at least 200 dpi, which does not show pixelation.

Figures are numbered sequentially and should be referred to in the text. If possible, integrate the reference into the sentence.

…, as shown in Figure #.
Figure # illustrates ….

Figures have either a title or a caption, with captions being more common in published work. Captions are below the figure, explain the data presented, and provide concise details on information in the figure that may be difficult to understand. The caption should be a maximum of three sentences. The title or caption must contain sufficient information to explain the figure without referencing the text.

Color in figures, images, and graphs

Color is readily available in word processors, spreadsheets, and graphics software. Used correctly, color highlights information and makes interpretation easier. However, the cost is three to five times higher for both students and publishers. Unlike high school and first-year textbooks, most advanced textbooks and scholarly articles are printed in grayscale. Alternatives to color include using different symbols, line styles, and shading to differentiate data in grayscale figures (see page 100).

Another problem with figures is that reproduction (photocopying) is not perfect. Subtle shades and colors are lost when reproduced. This is especially true for images. When using color, ensure the entire figure is interpretable when photocopied or printed in grayscale.

Figures

Common figures include flow charts, instrument schematics, and line drawings. They are created to illustrate a process, and can be created with varying amounts of detail, depending on the audience.

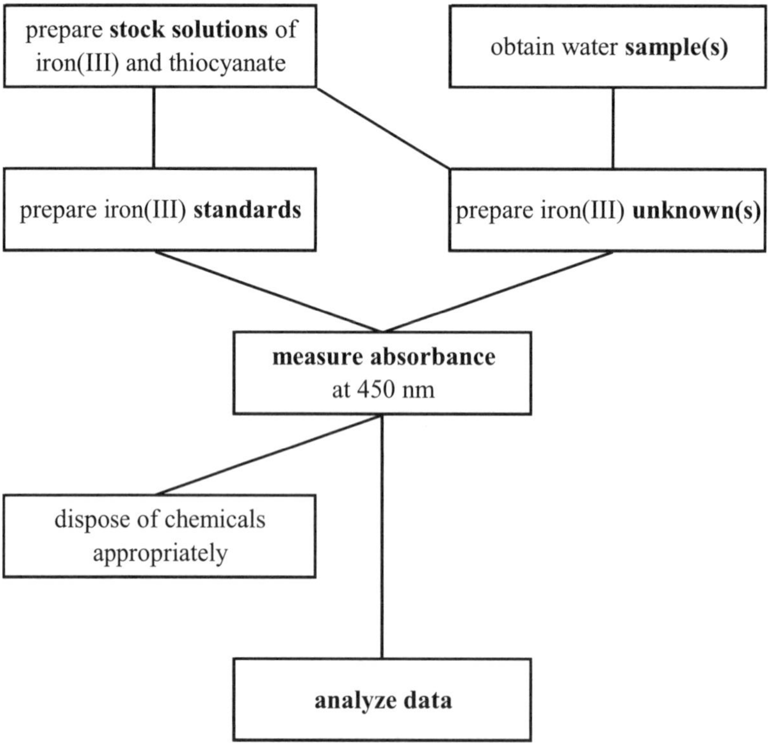

Figure 2.1 Flowchart summarizing the procedure for determining the iron concentration in water samples.

Figure 2.2 A line drawing showing the variation in the finch beaks on the Galapagos islands as observed and recorded by Charles Darwin.

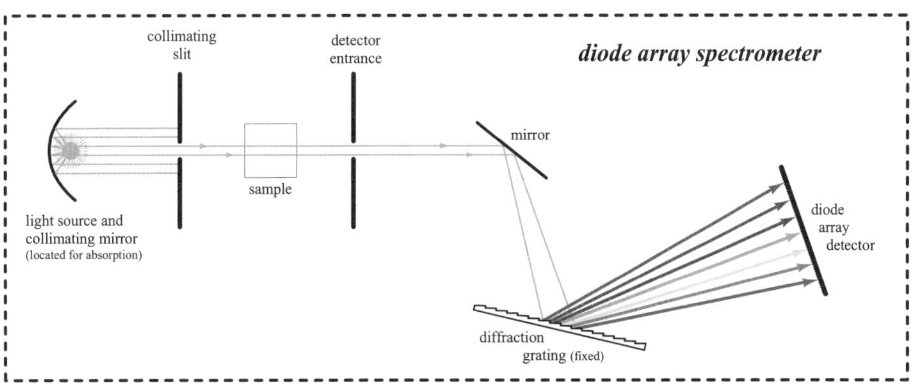

Figure 2.3 Schematic of a spectrometer configured for measuring absorption.

Images

The schematic in Figure 2.3 simplifies the instrumentation by reducing it to its fundamental components. Figures 2.4 and 2.5 are images of a diode array spectrometer and the instrumental setup, respectively. In Figure 2.5, the surrounding instrumentation is hidden so the image focuses on the instrument of interest.

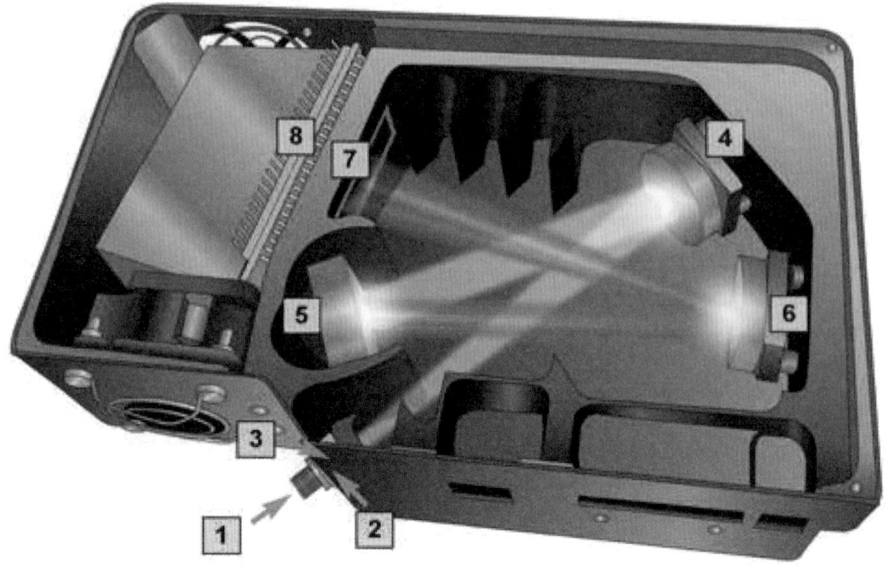

Figure 2.4 Realistic representation of an Ocean Optics QE 6500 diode array spectrometer showing the components and light path. Compared with Figure 2.3, ①, ② and ③ are the detector entrance and slits, ④ and ⑥ are mirrors, ⑤ is the diffraction grating, ⑦ is the optical filter, and ⑧ is the diode array detector. The source and sample are external to the spectrometer and not shown. (Used with permission.)

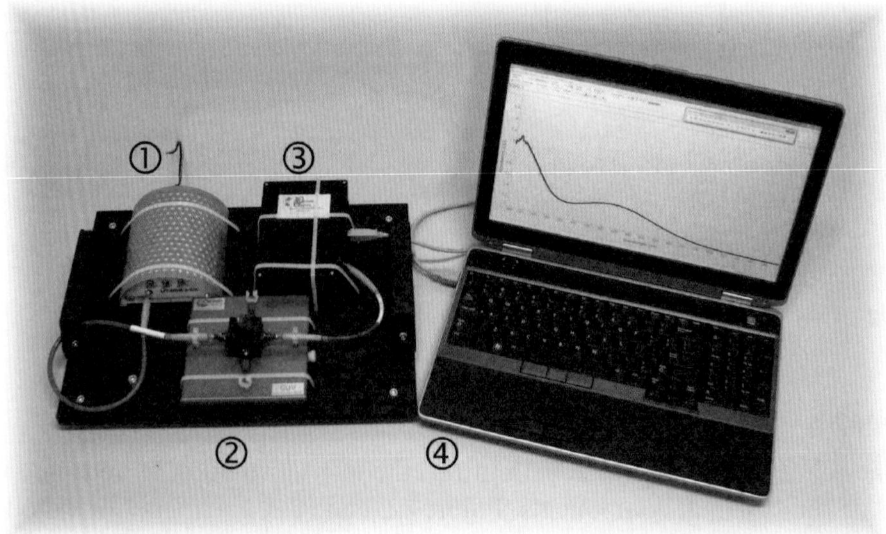

Figure 2.5 Image of the experimental setup for measuring absorbance. The light source, ①, sample, ②, and spectrometer, ③, are connected via fiber optic cables (blue). The spectrometer is controlled by a computer, ④, via a USB cable.

Figures 2.3, 2.4, and 2.5 present information on the same experimental apparatus, but the information conveyed in each figure is different. Figure 2.3 presents a simplified schematic. Figures 2.4 focuses on the detector. Figure 2.5 presents the actual experimental setup. The figure required in a document depends on the information to be conveyed.

Graphs

Graphs allow the reader to visualize data and identify trends in the data. You can find examples of quality graphs in *Communicating Science*, in textbooks, and in published scholarly articles.

Figure 2.6 presents the Cartesian axis with possible graphs that you may be expected to make. Observe that graphs do not need to include {0, 0}.

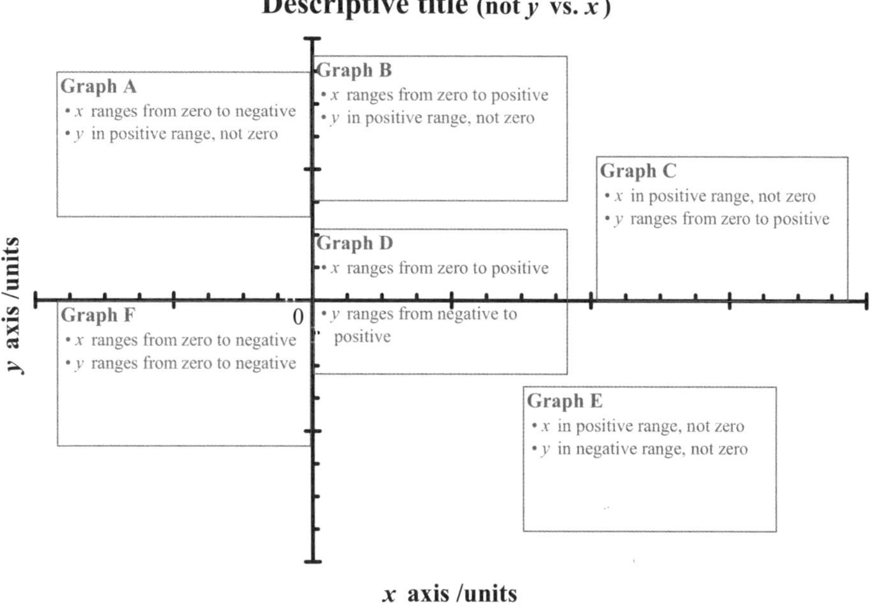

Figure 2.6 Cartesian coordinate system showing some possible graphs, each of which are a subset of the coordinate system.

You are probably familiar with the normal numbering of axes. Additionally, axes may also be logarithmic, which is sometimes confusing to plot and read. Figure 2.7 illustrates the numbering of a logarithmic y axis. (Both the x and y axes can be logarithmic.)

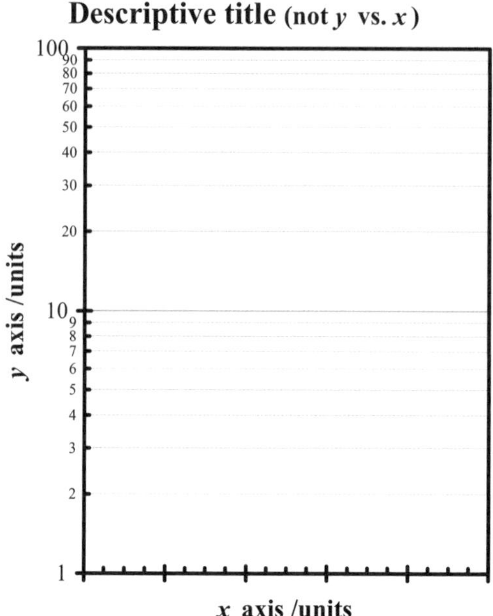

Figure 2.7 A graph with a logarithmic y axis, illustrating how the axis is numbered.

Features of a quality scientific graph

Plotting data:

- The x-axis is the controlled (independent, manipulated) variable; the y-axis is the dependent (responding) variable.
- The axes ranges should be set so that the plotted data occupies most of the graph area while still having reasonable ranges.
- Adjust the axes scales so there are between four and six numbers on the axes, and the tick marks are a reasonable integer or fractional interval.
- The numerical values on the axes should be to a minimum number of significant digits (not the number of significant digits in the data points).*
- Individual data points are represented with markers: ■, ●, ♦, □, ○, …
    - There should *not* be a line joining the data points.
    - *Exception*: if there are so many data points that they cannot be individually seen, such as the spectrum in Figure 2.8, it is

* In advanced statistics courses, you will learn that the number of significant digits (the precision) in the data points translates to a weighting of the data. The uncertainty in the slope and intercept are calculated statistically based on the scatter in the data.

permissible to use a line to join the data points and not display the individual data points.

- Best-fit lines (trendlines), if used, should be added by the software.
- Include a legend if you have more than one data set on a graph.
- Do not plot gridlines unless you are extracting data directly from the graph.

Labeling the graph:

- The axes labels must identify what is being plotted and the units.
- Use either a title or caption — not both.
- The title must provide additional information about the graph: the system studied, the object of analysis, etc. *Do not repeat the axes labels!*
- Adjust the fonts, colors, and line thicknesses so they are legible in the final work.

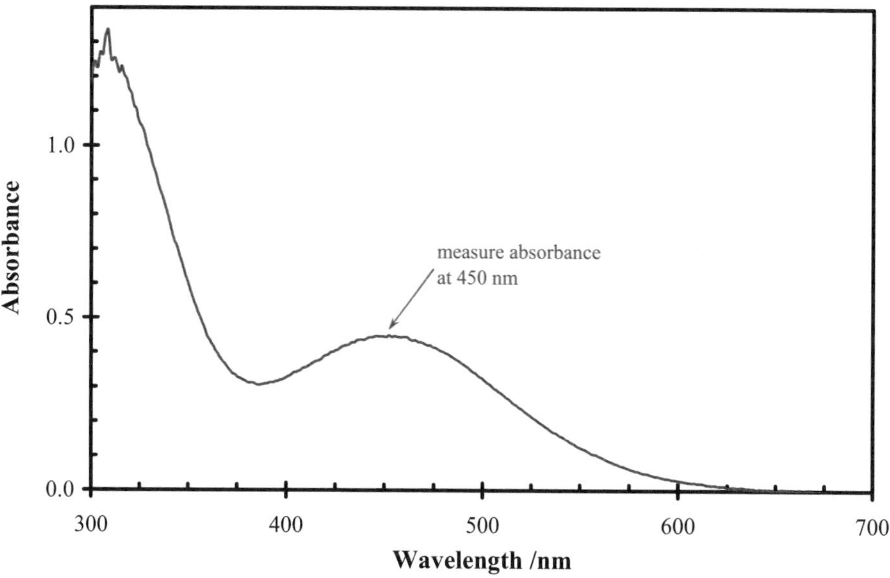

Figure 2.8 Visible absorption spectrum of iron(III) thiocyanate, $Fe(SCN)^{2+}$.

There are 2000 data points from 300 nm to 700 nm in Figure 2.8. Instead of plotting every data point, a line is drawn between the data points. Also observe that the *x*-axis does not start at zero.

The wavelength plotted on the x-axis has units of nanometers. These units are transferred to the axis label using the mathematical procedure presented previously.

$$\text{wavelength} = 500\text{ nm} \quad \Rightarrow \quad \frac{\text{wavelength}}{\text{nm}} = 500$$

$$\underbrace{\text{wavelength /nm}}_{\text{axis label}} = \underbrace{500}_{\text{plotted value}} \qquad \mathbf{2.4}$$

Table 2.2 (page 89) contains the concentration and absorbance data of the iron(III) thiocyanate standards and unknown. Plotting the standards produces the calibration curve plotted in Figure 2.9.

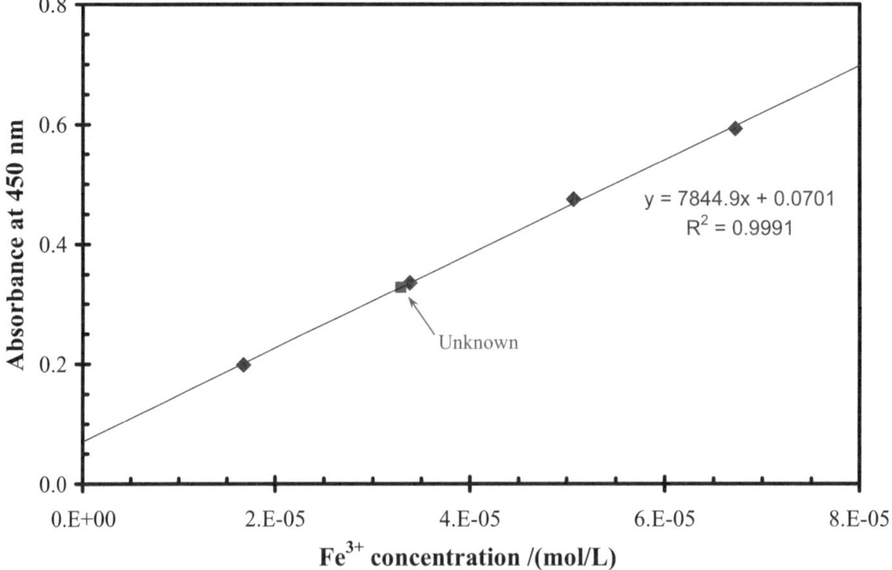

Figure 2.9 Iron(III) thiocyanate calibration curve for the determination of iron in the water sample.

The trendline in Figure 2.9 returns the equation of the line and the correlation coefficient, r^2.* The computer-generated equation does not account for significant figures or units, but does allow the unknown concentration to be calculated. Proper units are added below.

* Consult a statistics resource to learn more about statistical analysis methods.

$$y = 7844.9\,x + 0.0701 \quad \Rightarrow \quad A_{450\text{ nm}} = 7844.9\,\frac{\text{L}}{\text{mol}}\left[\text{Fe}^{3+}\right] + 0.0701$$

$$A_{450\text{ nm}} + 0.0701 = 7844.9\,\frac{\text{L}}{\text{mol}}\left[\text{Fe}^{3+}\right]$$

$$\left[\text{Fe}^{3+}\right] = \frac{A_{450\text{ nm}} + 0.0701}{7844.9\,\frac{\text{L}}{\text{mol}}}$$

The unknown concentration is determined using the measured absorbance.

$$\left[\text{Fe}^{3+}\right]_{\text{ukn}} = \frac{0.329 + 0.0701}{7844.9\,\frac{\text{L}}{\text{mol}}} = \boxed{3.30\cdot 10^{-5}\,\frac{\text{mol}}{\text{L}}} \qquad \textbf{2.5}$$

Significant figures in the original measurements dictate the significant figures in the unknown iron(III) concentration.*

Hand-drawn graphs

Drawing graphs by hand is the *best* way to learn how to create a quality graph that conveys the most information possible. Once mastered by hand, it is much easier to create a quality graph on a computer.

Figures 2.8 and 2.9 are computer-drawn. Figure 2.10 is a hand-drawn version of Figure 2.9.

When preparing graphs by hand,

- Use 1 mm grid graph paper to assist in pinpointing the data points and slope.
- Watch out for data that is not linear! Non-linear data has systematic deviations above and below a "best fit" straight line.
- Do not draw a line connecting every data point.
    - If the data is linear, use a ruler to draw a best-fit line. Position the ruler so that the data points are evenly scattered above and below the line.
    - If the data is non linear, draw a smooth line by hand without a ruler so that the data points are evenly scattered above and below the line.

* This is an approximation. Advanced statistics courses present a method for calculating the uncertainty of values extracted from graphical analyses.

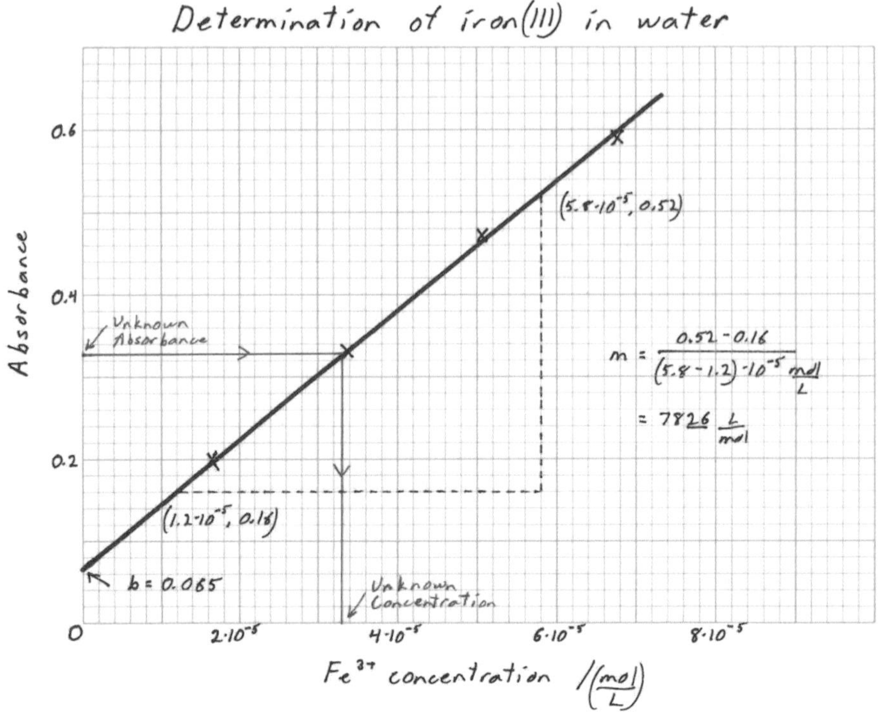

Figure 2.10 Hand-drawn graph showing the data points, trendline, gridlines, slope and intercept calculations, and how an unknown sample was analyzed on the graph. The slope and intercept may differ slightly because of errors in estimating the best-fit line by hand.

Plotting multiple data sets

It is common to plot multiple data sets on the same graph to show the similarities and differences between the data sets. Take care when selecting colors because some colors do not print or photocopy well. As a rule, avoid using yellow and light (pastel) colors, which are difficult to see. Other colors may have the same grayscale shade, making it difficult to identify the data sets.

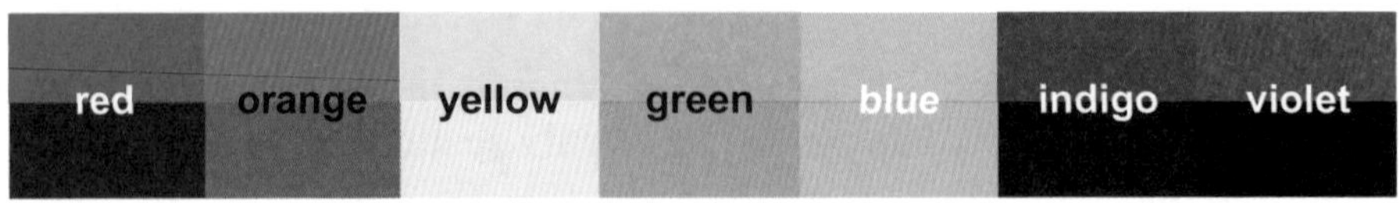

Figure 2.11 Select colors and their grayscale equivalents.

To assist readers, you should use two methods to distinguish the data sets. Possible methods include

- colored lines, selected so that their grayscale colors are different
- labeling each line with a number, letter, label, or obvious identifier
- if there are few data points, use different data-point markers that are the same color as the line: ■, ●, ♦, □, ○, …
- if there are many data points, use different lines styles

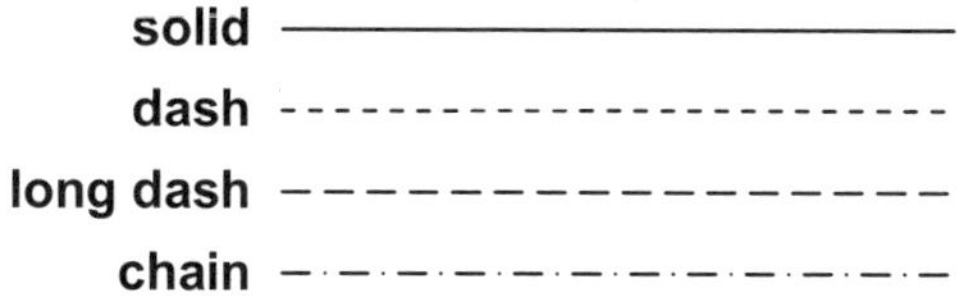

Figure 2.12 Selected line styles. Unless one line is more important, use the same line width for all lines. Section B.2 (page 276) provides information on creating publication-quality graphs.

To identify the data set, include a legend either on the graph or in the caption.

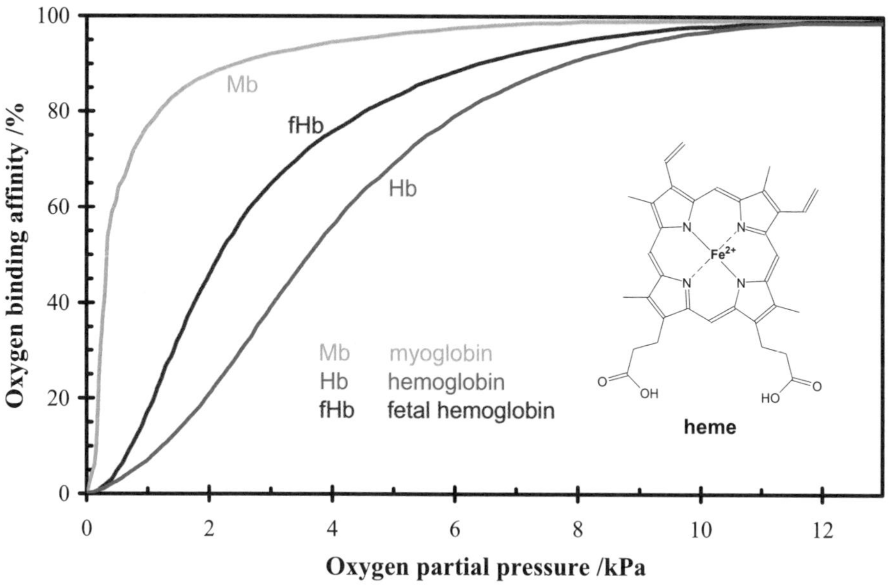

Figure 2.13 Binding affinity of oxygen to heme when heme is in different proteins.

Graphing non-linear data

Graph non-linear data like linear data. While linear regression is not possible, it is possible to fit a mathematical equation to the data and plot that fit. (See the section on using Excel® Solver in Appendix B.)

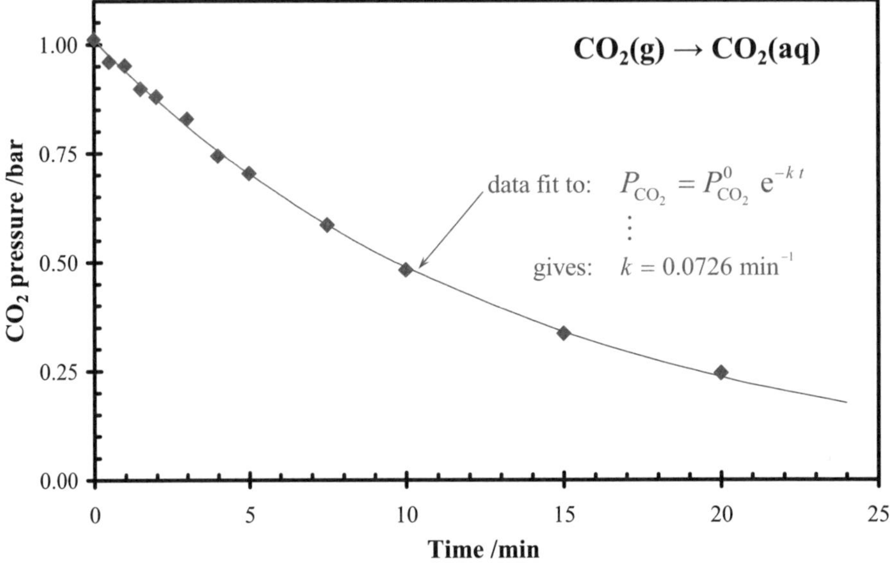

Figure 2.14 Dissolution of carbon dioxide in water occurs as a first-order reaction with a rate constant of $0.0726\ min^{-1}$.

Scatter plots

The data presented in Figures 2.9, 2.10, 2.14 show little scatter, so only a few data points are necessary to determine the relationship between the data. However, scatter is common in some experiments, and more data is required to determine the relationship. When preparing scatter plots, choose a dot size that makes each data point visible.

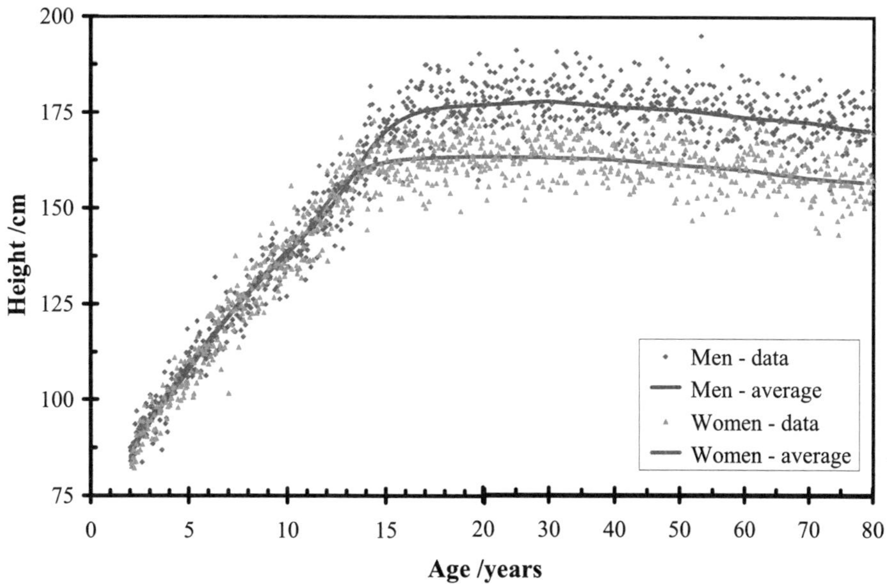

Figure 2.15 Height:age relationship for men and women aged 2 to 80.

Source: data on 2–20 year-olds adapted from Centers for Disease Control and Prevention. Data Table of Stature-for-age Charts (internet). Accessed 17 February 2013. Available from http:// www.cdc.gov/growthcharts/html_charts/statage.htm

Source: data on 20–80 year-olds adapted from National Center for Health Statistics. June 1965. Weight, Height, and selected Body Dimensions of Adults. Series 8 (11).

Bar graphs

In some cases, numerical data can be grouped into categories that are not quantitative (i.e., not numerical). To represent this data graphically, a bar graph plots the data in each category as vertical or horizontal bars, evenly spaced along the axis. Numerical data with a regular interval can also be displayed in a bar graph, but only if there are fewer than ten data points.

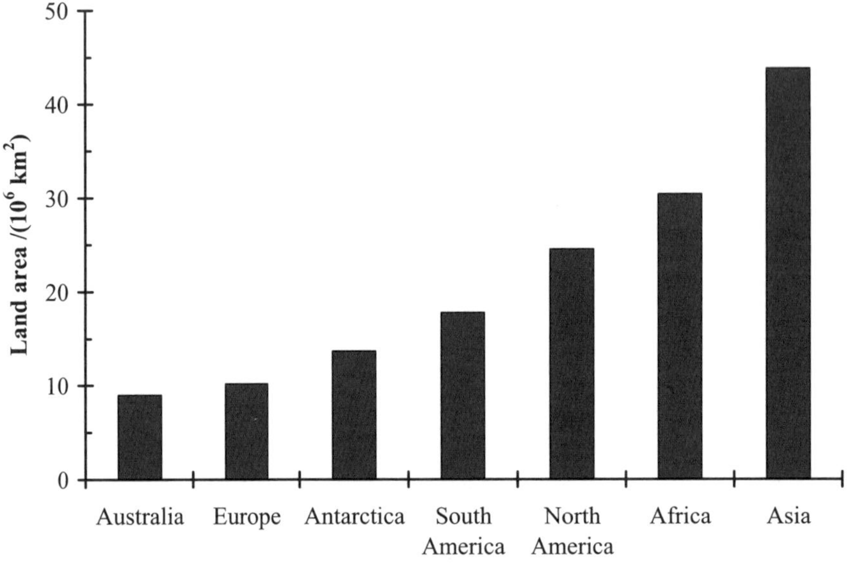

Figure 2.16 The surface area of the continents.

Source: Wikipedia. Continent (internet). Accessed 17 February 2013. Available from http://en.wikipedia.org/wiki/Continent

Figures 2.17 and 2.18 present two ways multiple data from the same category may be presented. The *grouped bar chart* in Figure 2.17 presents the data as separate bars, making it easy to determine the individual contributions and how they vary across the data sets. (Figure 2.21 presents the same data as Figure 2.17, but as a 3D bar graph.) The *stacked bar chart* in Figure 2.18 presents the total power produced, but makes it more challenging to determine the individual contributions.

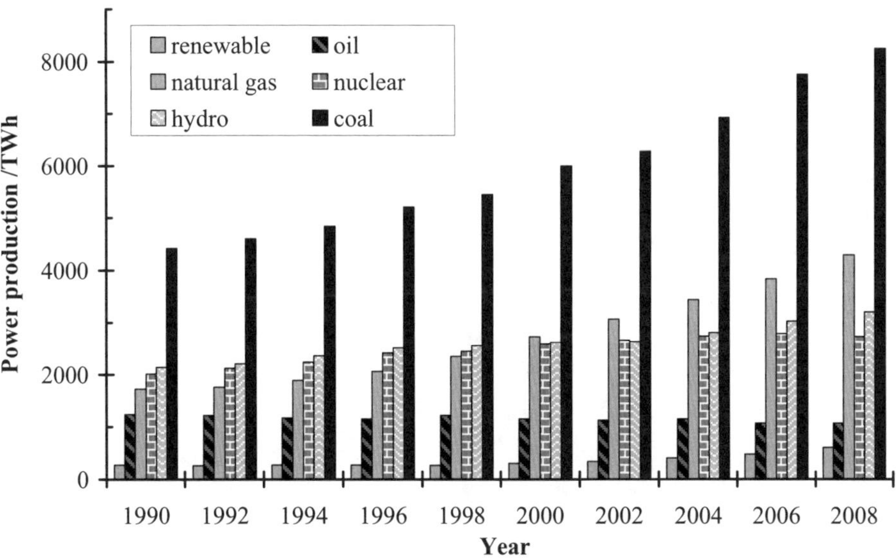

Figure 2.17 Global sources of power during the period 1990 – 2008.

Source: data from Statens energimyndighet. 2010 Energy in Sweden - facts and figures 2010.Table 48.
Note: I added patterning because the adjacent colors are similar if printed in grayscale.

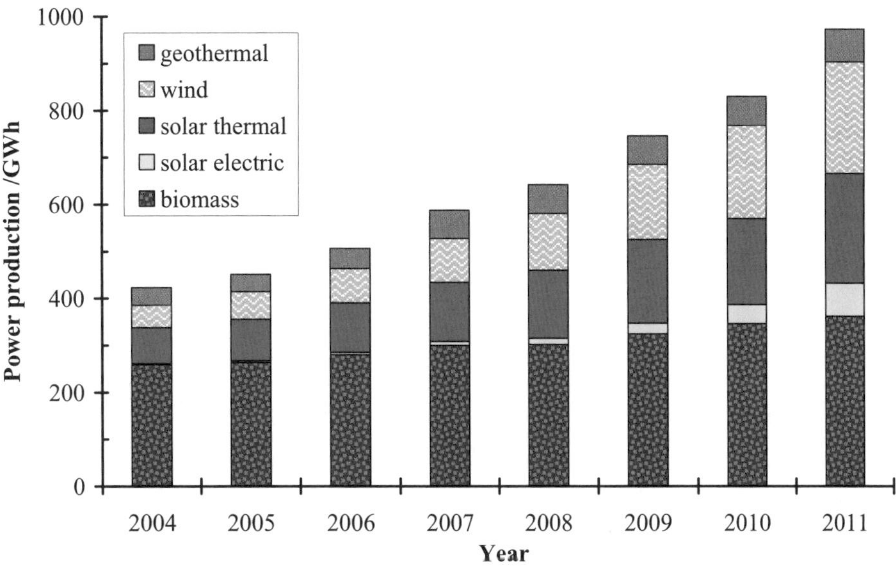

Figure 2.18 Sources of renewable power during the period 2004 – 2011.

Source: data from Renewable Energy Network for the 21st Century (REN21). Renewables <year> Global Status Report, (Table R1 in the reports from 2005 – 2012) (internet). Available from www.REN21.net
Note: I added patterning because the adjacent colors are similar if printed in grayscale.

Pie charts

When all components contribute more than two percent each to the whole and your intent is to convey a general impression — not to extract quantitative data — it is possible to present the data as a *pie chart* (*circle chart*). If quantitative information must be conveyed, present the data in a table or bar chart, or include the quantitative data in the pie chart as shown in Figure 2.19.

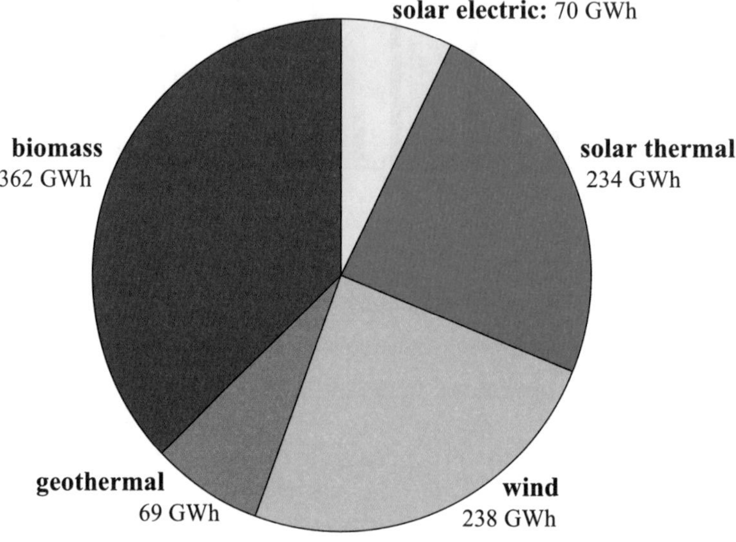

Figure 2.19 2011 Renewable power production. Not shown is tidal/wave power, which produced 0.5 GWh.

Source: data from Renewable Energy Network for the 21st Century (REN21). 2012. Renewables 2012 Global Status Report (internet). Table R1. Available from www.REN21.net

3D Graphs

Current computing technology makes it possible to create 3D surface and bar graphs. In these graphs, the *x*- and *y*-axes represent independent variables, while the *z*-axis represents the dependent variable.

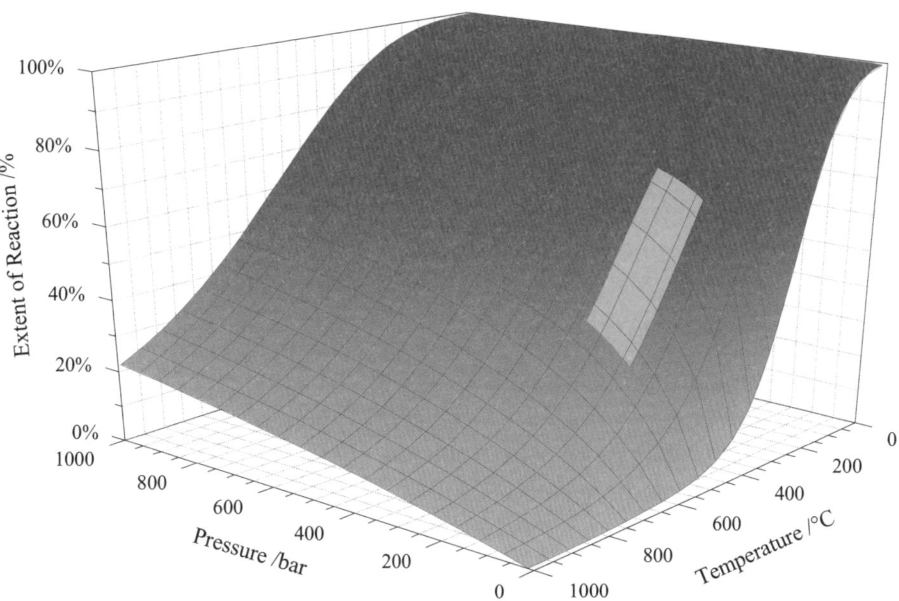

Figure 2.20 The extent of reaction of ammonia synthesis: $N_2(g) + 3H_2(g) \rightleftharpoons 2\ NH_3(g)$

While the reaction favors products at low temperatures, the reaction is kinetically hindered at low temperatures. The colored patch indicates the conditions used industrially to synthesize ammonia, which is a compromise between yield and reaction rate.

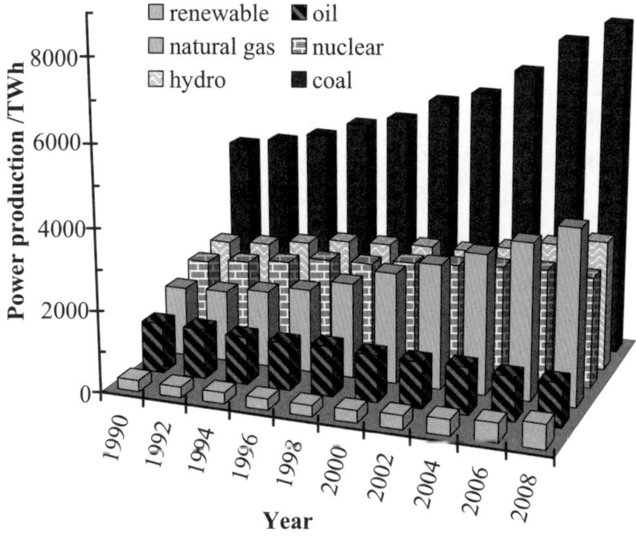

Figure 2.21 Sources of global power during the period 1990 – 2008.

Source: data from Statens energimyndighet. 2010 Energy in Sweden - facts and figures 2010. Table 48.

Note: I added patterning because the adjacent colors are similar if printed in grayscale.

2.7 An introduction to statistical analysis

This section both introduces statistical data analysis tools and illustrates the use of equations, figures, and tables in scientific documents.

When collecting data, scientists often take multiple measurements of some samples. Statistical analysis of those measurements provides information on the uncertainty in the data.

Rejecting data (Q-test)

Given n measurements of a sample, if one data point is distant from the other data points — an *outlier* — the Q-test may be used to determine if the outlier can statistically be rejected. Q_{calc} is calculated from the experimental data and measured against Q_{tab}. Q_{tab} is a statistical measure of the probability that Q would happen — that one data point could be distant from the remainder — by chance at a given confidence level.

$$Q_{calc} = \frac{gap}{range} = \frac{|suspect - nearest|}{largest - smallest} \qquad \textbf{2.6}$$

Table 2.4 tabulates Q_{tab} for common confidence levels. If $Q_{calc} > Q_{tab}$, the outlier can be rejected at the specified confidence level.

Table 2.4 Values of Q_{tab} for n measurements at common confidence levels.

n	68%	90%	95%	98%	99%
3	0.822	0.941	0.970	0.988	0.994
4	0.603	0.765	0.829	0.889	0.926
5	0.488	0.642	0.710	0.780	0.821
6	0.421	0.560	0.625	0.698	0.740
7	0.375	0.507	0.568	0.637	0.680
8	0.343	0.468	0.526	0.590	0.634
9	0.319	0.437	0.493	0.555	0.598
10	0.299	0.412	0.466	0.527	0.568
12	0.271	0.375	0.425	0.480	0.518
14	0.250	0.350	0.397	0.447	0.483
16	0.234	0.329	0.376	0.422	0.460
18	0.223	0.314	0.358	0.408	0.438
20	0.213	0.300	0.343	0.392	0.420
⋮	⋮	⋮	⋮	⋮	⋮

Data transformation

In most cases, you must mathematically transform your raw data into the quantity of interest, such as converting from absorbance to concentration. It is important to convert each data point to a final value before statistically analyzing the data. Why? Because non-linear mathematical operations (square root, power, logarithm, etc.) skew the distribution of the results. *There is a difference* if each data point is transformed to the final value and then averaged *or* if the raw data is averaged and then transformed to the final value.

Average

Average is a general term and can be the mean, median, or mode for n observations of a sample.

The *mean*, $\bar{x}$, is calculated from the formula:

$$\bar{x} = \frac{1}{n}\sum_{i=1}^{n} x_i \qquad \textbf{2.7}$$

The *median* is the middle data point after the data is sorted in ascending or descending order. If there are an even number of data points, the median is the mean of the center two data points.

The *mode* is the most frequently observed value, but can only be determined in large data sets.

Standard deviation

Standard deviation, s, is a measure of the precision of a single data point in a set of data points obtained by taking replicate measurements of a sample. The standard deviation may be thought of as the range in which we expect the next observation to be found with a certain confidence.

$$s = \sqrt{\frac{\sum_{i=1}^{n}\left(x_i - \bar{x}\right)^2}{n-1}} \qquad \textbf{2.8}$$

Degrees of freedom

The concept of degrees of freedom (*d.f.*) refers to the number of independent values in a calculation. Every statistical calculation reduces the degrees of freedom by one.

Given n measurements, there are initially n degrees of freedom.

- Calculating the average removes one degree of freedom.
- Calculating the standard deviation removes another degree of freedom. (Note the $n - 1$ in Equation 2.8.)

Encompassing the true value: confidence intervals

In addition to calculating the average and standard deviation, it is possible to calculate the range that encompasses the true value with a statistical confidence. Multiplying the standard deviation by a factor t (often called Student's t) determines the *confidence interval* (Δx, uncertainty) of the average.

$$\Delta x = \frac{t\,s}{\sqrt{n}} \qquad \textbf{2.9}$$

The true value, μ, is within the confidence interval at the stated confidence level.

$$\mu = \bar{x} \pm \Delta x \qquad \textbf{2.10}$$

Table 2.5 lists two-tailed t-values for varying degrees of freedom at common confidence levels. *Two-tailed* means that the true value could be greater or less than the average.

The true value in Equation 2.10 may be reported in two ways:

- reporting the confidence level: 90 % 95 % 98 %
    - The ***confidence level*** is the probability, as a percentage, that the true answer is within the stated range.
- reporting the p-value: $p < 0.10$ $p < 0.05$ $p < 0.02$
    - The ***p-value*** is the probability, as a fraction, that the true answer is outside the stated range.

The confidence level and p-value are related:

$$\frac{\textit{confidence interval}}{100\,\%} + p\text{-value} = 1.00 \qquad \textbf{2.11}$$

> Polls and surveys often contain a statement like, "The poll/survey is accurate to within three percentage points 19 times out of 20." Statistically, this statement translates to an uncertainty of ± 3 % at the 95 % confidence level. (19/20 · 100 % = 95 %)

Table 2.5 Two-tailed t-values for varying degrees of freedom at common confidence levels.

d.f.	Two-tailed *t*-test				
	68%	**90%**	**95%**	**98%**	**99%**
2	1.312	2.920	4.303	6.965	9.925
3	1.189	2.353	3.182	4.541	5.841
4	1.134	2.132	2.776	3.747	4.604
5	1.104	2.015	2.571	3.365	4.032
6	1.084	1.943	2.447	3.143	3.707
7	1.070	1.895	2.365	2.998	3.499
8	1.060	1.860	2.306	2.896	3.355
9	1.053	1.833	2.262	2.821	3.250
10	1.046	1.812	2.228	2.764	3.169
12	1.037	1.782	2.179	2.681	3.055
14	1.031	1.761	2.145	2.624	2.977
16	1.026	1.746	2.120	2.583	2.921
18	1.023	1.734	2.101	2.552	2.878
20	1.020	1.725	2.086	2.528	2.845
⋮	⋮	⋮	⋮	⋮	⋮

The following example calculates the copper concentration in a sample of brass at the 95 % confidence level.

Five pieces of brass were analyzed and the percentage of copper in each measured as 93.42 %, 93.86 %, 92.18 %, 93.14 %, and 93.60 % by mass.

The data is visualized at right. 92.18 % appears to be an outlier. Applying the Q-test (Equation 2.6) to the data gives

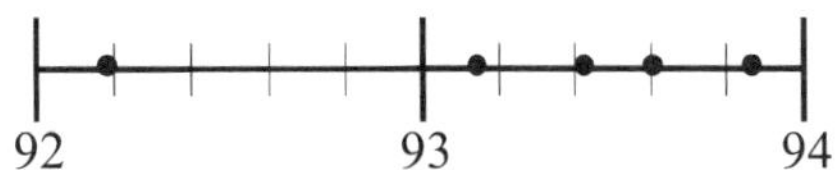

$$Q_{\text{calc}} = \frac{gap}{range} = \frac{|92.18\,\% - 93.14\,\%|}{93.86\,\% - 92.18\,\%} = 0.571$$

From Table 2.4, $Q_{\text{tab}}(95\ \%, n = 5) = 0.710$. Since $Q_{\text{calc}} < Q_{\text{tab}}$, the value cannot be rejected at the 95 % confidence level.

Keeping all the measurements, the mean is 93.36 % (Equation 2.7) and the standard deviation is 0.417 %, calculated using Equation 2.8. Calculating the mean uses one degree of freedom. From Table 2.5, $t(95\ \%, d.f. = 4) = 2.776$, and the uncertainty, Δx, is calculated using Equation 2.9:

$$\Delta x = \frac{t\,s}{\sqrt{n}} = \frac{2.776 \cdot 0.417\,\%}{\sqrt{5}} = 0.518\,\%$$

Uncertainties are typically reduced to one significant digit, so the result would be reported as,

> "The concentration of copper in the brass is (93.4 ± 0.5) % by mass at the 95 % confidence level."

or

> "The concentration of copper in the brass is (93.4 ± 0.5) % ($p < 0.05$) by mass."

Sources of uncertainty (ANOVA)

The *variance* is the standard deviation squared.

$$V = s^2 \qquad \textbf{2.12}$$

Variance is additive for normal (Gaussian) distributions, making it possible to determine the magnitude of different sources of uncertainty. This analysis is often called an ANalysis Of VAriance (ANOVA).

Figure 2.22 shows how to analyze a sample to determine how sampling, preparation, and analysis contribute to the total uncertainty. The variance in **A** is due to analysis only; the variance in **B** is due to analysis and preparation; the variance in **C** is the total variance of all processes. The additive nature of variance gives

$$V_{\text{total}} = V_{\text{sampling}} + V_{\text{preparation}} + V_{\text{analysis}} \qquad \textbf{2.13}$$

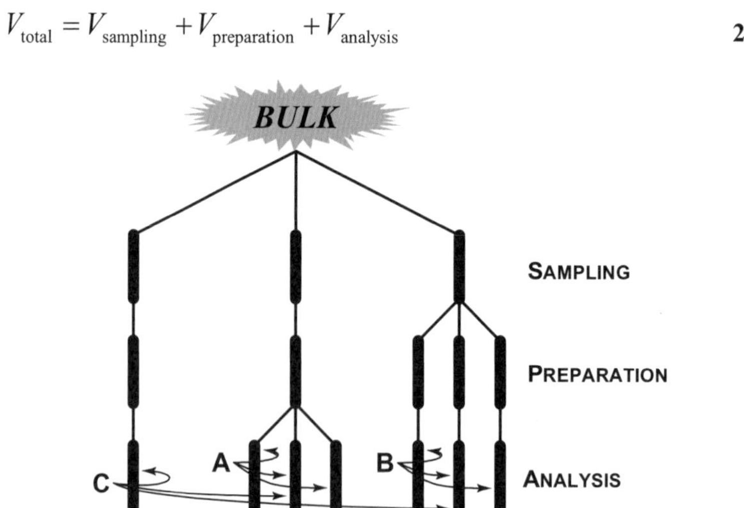

Figure 2.22 Sample flow-chart for conducting an ANOVA.

Propagation of uncertainty

You have probably learned about how significant digits are carried through mathematical transformations (calculations). The significant digits method is a simple, but imprecise, method of estimating the uncertainty in the final value. In statistics, both the data and uncertainty are transformed; this process is called ***propagation of uncertainty*** (***propagation of error***). Table 2.6 lists the functions for propagating uncertainty (reported as the standard deviation) through common mathematical operations.*

Table 2.6 Functions for propagating uncertainty through common mathematical operations.

Operation	Uncertainty	Operation	Uncertainty (base e)	Uncertainty (base 10)
$z = x + y$ $z = x - y$	$s_z^2 = s_x^2 + s_y^2$	$z = \ln(x)$ $z = \log(x)$	$s_z = \frac{s_x}{x}$	$s_z = \frac{1}{\ln(10)} \frac{s_x}{x}$
$z = x \cdot y$ $z = x/y$	$\left(\frac{s_z}{z}\right)^2 \approx \left(\frac{s_x}{x}\right)^2 + \left(\frac{s_y}{y}\right)^2$	$z = e^x$ $z = 10^x$	$\frac{s_z}{z} = s_x$	$\frac{s_z}{z} = \ln(10) s_x$
$z = x^a$ $= x \cdot x \cdot \ldots$	$\frac{s_z}{z} = a \frac{s_x}{x}$		↑ base e	↑ base 10

* Equations where the uncertainty in z is approximate have ignored covariance between x and y. This is a common approximation, but has varying validity.

Additional resources …

… *on communicating science*

AIP Style Manual (internet). 4th ed. New York: American Institute of Physics; 1990. Available from http://www.aip.org/pubservs/style/4thed/AIP_Style_4thed.pdf

Chandrasekhar R. How to Write a Thesis: A Working Guide [essay; internet]. Crawley (Australia): University of Western Australia; 2002. Available from http://ciips.ee.uwa.edu.au/pub/HowToWriteAThesis.pdf

Cohen ER, Cvitas T, Frey JG, Holstrm B. Kuchitsu K, Marquardt R, Mills I, Pavese F, Quack M, Stohner J, Strauss HL, Takami M, Thor AJ, (editors). Quantities, Units and Symbols in Physical Chemistry. 3rd ed. Cambridge: Royal Society of Chemistry; 2007.

Council of Science Editors. Scientific Style and Format: The CSE Manual for Authors, Editors, and Publishers. 7th ed. Reston, VA: Rockefeller University Press; 2006.

Lim KF. The Chemistry Style Manual. 2nd ed. Victoria (AU): Deakin University Press; 2003.

McCaskill M. Grammar, Punctuation, and Capitalization: A Handbook for Technical Writers and Editors. NASA SP-7084. Hampton, VA: Langley Research Center; 1998.

Paradish J, Zimmerman, M. The MIT guide to science and engineering communication. 2nd ed. Cambridge, MA: MIT Press; 2002.

Chapter 3. Fundamentals of learning

In order to prepare quality documents and give quality presentations, it is valuable to have an understanding of how people learn.

Table 3.1 Common and scientific definitions of *knowledge* and *learning*.

Common definition	Scientific definition
Knowledge consists of the accumulated facts and impressions we retain from previous experiences.	***Knowledge*** exists as neural networks in the brain.
Learning is the process of increasing knowledge.	***Learning*** occurs when new connections are formed between neurons, expanding the neural network.

Importantly, knowledge is gained and shared through the communication of information. Clear, coherent, concise, and precise communication facilitates knowledge transfer and learning by others.

The ***constructivist theory*** of learning proposes that learning occurs when new experiences and new information integrates into our pre-existing knowledge framework, thus expanding the framework.* The constructivist theory originated as an empirical proposal in the 1600s. Subsequent research into biophysical processes within the brain has provided a theoretical foundation for the constructivist theory.

> *Tell me and I will forget; show me and I may remember; involve me and I will understand.* — Chinese proverb

Brain function

Research into brain function is a very active field. With the development of real-time MRI, researchers have a new and powerful tool to explore structure-function and activity-function relationships within the brain.

The brain is composed primarily of neurons, which are cells responsible for processing, transmitting, and storing information. Neurons contain

- *dendrites*, which receive signals from other neurons
- an *axon*, which propagates signals within a neuron
- *synaptic terminals*, which transmit signals to other neurons

* The constructivist theory is complex and nuanced. Only the fundamental premise is presented here. The Wikipedia entry on *constructivism* and the references cited therein are excellent starting points to learning about the constructivist theory.

The *synapse* is the region between a synaptic terminal on one neuron and a dendrite on another. It is still not well understood how a neuron knows which neuron it is receiving a signal from or how it selects which neuron to transmit the signal to. It is known that *learning* is the formation of connections between neurons, which leads to the development and expansion of *neural networks*.

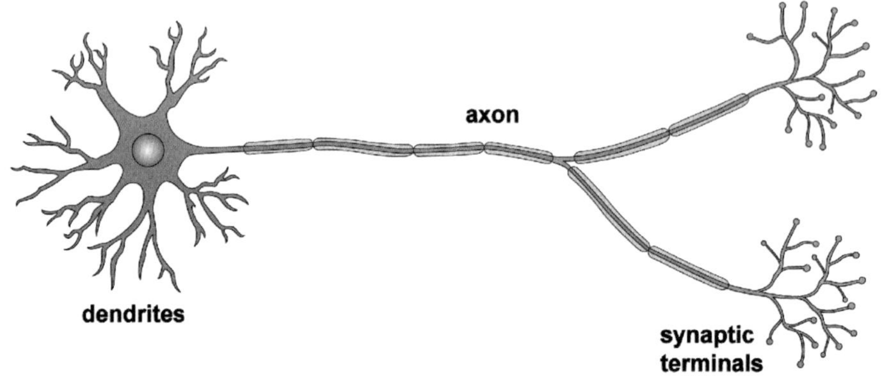

Figure 3.1 Line drawing of the major components of a neuron.

Source: Image ID 29889581 © MrHighSky | DreamsTime.com

Kolb's learning model

Dr. David Kolb is an educational theorist at Case Western Reserve University. In 1985, he developed a model for how people learn. Kolb proposes that a person cycles through four stages when learning:

- abstract hypothesis (planning)
- active testing (experimentation)
- concrete experience (observation)
- reflective observation (review and analysis)

These learning stages are illustrated in Figure 3.2.

Dr. James Zull, a biologist also at Case Western Reserve University, discovered a link between Kolb's model and the regions and pathways in the brain. Zull's work, also presented in Figure 3.2, provides a biophysical foundation for Kolb's model.

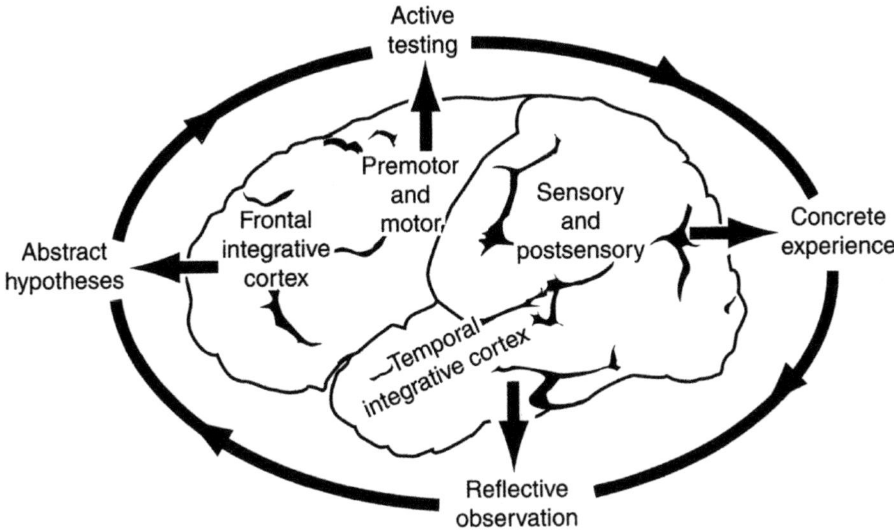

Figure 3.2 The relationship between Kolb's learning model and functional regions in the brain. (Used with permission from Dr. James Zull.)

The regions of the brain have functions that match the learning stages proposed by Kolb.

- The *frontal cortex* is responsible for high-level reasoning.
- The *premotor and motor* region is responsible for coordinating and controlling the body.
- The *sensory and post-sensory* region receives sensory input from the senses and processes that information to develop an understanding of the surroundings.
- The *temporal cortex* is responsible for long-term memory storage.

Kolb also proposed learning styles as the transitions between the stages.

- *converging*: abstract hypothesis → active testing
- *accommodating*: active testing → concrete experience
- *diverging*: concrete experience → reflective observation
- *assimilating*: reflective observation → abstract hypothesis

People engaged in *convergent* learning apply theories to solve problems and find solutions to practical issues. (Persons who prefer this learning style commonly have careers in engineering, medicine, economics, and applied science.)

People engaged in *accommodative* learning adapt themselves to their immediate circumstances. They rely primarily on intuition instead of logic to solve problems. (Common careers include management, education, sales, and nursing.)

People engaged in *divergent* learning look at things from different perspectives. They watch rather than do and tend to gather information and generate ideas (brainstorm) to solve a problem. (Common careers include social work, acting, literature, and journalism.)

People engaged in *assimilative* learning take a concise, logical approach to solving a problem. They take information and organize it logically. They focus on ideas and abstract concepts. (Common careers include science, mathematics, and law.)

Figure 3.3 illustrates the four learning stages and four learning styles. Figure 3.3 is also the graph used by people determining their preferred learning stages and styles using Kolb's Learning Styles Assessment.

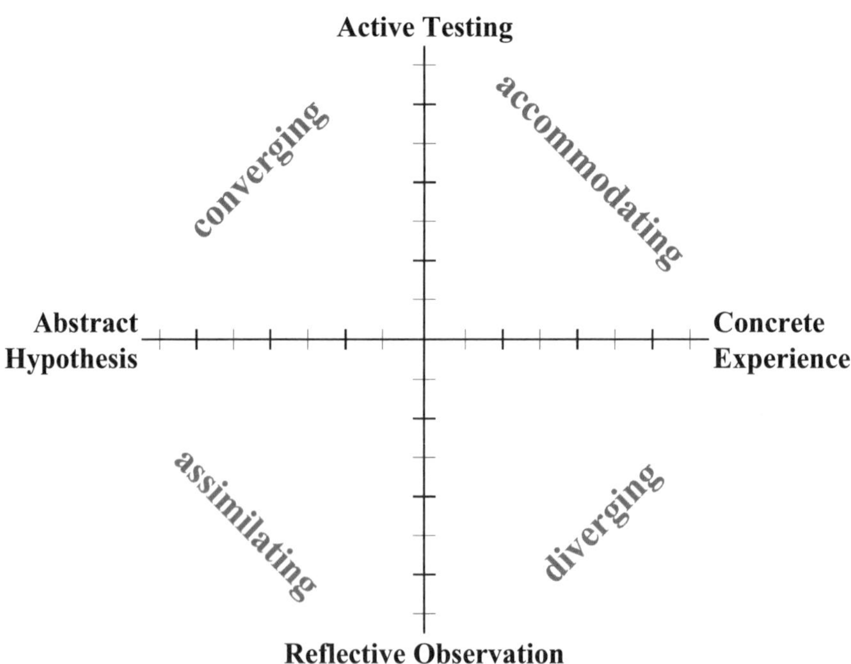

Figure 3.3 The four learning stages and four learning styles proposed by Kolb.

> The previous page lists careers of people who prefer a given learning style, but these people must be functional in all the other learning styles to be good in their career. *Practice* is required to become comfortable in all four learning styles. This practice can be in the form of a hobby: scientists who exercise and socialize develop their accommodative and divergent learning styles.
>
> Adept learners are comfortable in all four learning stages.
>
> Weakness at any stage introduces barriers to learning.

Learning

At birth, the brain has rudimentary neural networks to perform basic functions: breathing and circulation. Everything else is learned: muscle control, language, social interaction, etc. For example, an infant has very limited voluntary muscle control. Through random creation of neural pathways, an infant learns to suckle and move their fingers, arms, legs, and other parts of their body. If the neural pathway leads to a productive motion (suckling to receive milk, for example), the motion is repeated and the neural pathway reinforced. If the neural pathway is non-productive (a movement that causes pain, for example), the pathway is not used again and degenerates. As the neural pathways expand, the baby learns increasingly complex tasks: grasping, crawling, walking, eating, talking, etc. These abilities are refined through repeated use. (For example, a dancer or martial artist improves with practice.)

To see the Kolb cycle applied, envision an infant wanting to grab a toy:

1. The infant hypothesizes that, by extending their arm, they can reach and grab the toy.
2. The motor cortex is activated to extend the arm.
3. The sensory cortex receives input from the eyes and hand; the eyes see the hand extend and miss the toy; the nerves in the hand signal that they did not grab the toy.
4. The temporal cortex records that the hand did not extend far enough, missing the toy.
5. The infant hypothesizes that by extending their hand further, they can reach the toy.
6. … the cycle repeats …

Learning is a process. The first exposure to new information or to a new action begins neural network development. Repeated use of the neural network strengthens the pathways. The more the pathways are used, the better the information or action is learned. Use also creates pathways to related information that further expands understanding. If unused, the pathway degenerates and knowledge is lost.

Consider the following:

- Watching baseball games does not mean you can hit a baseball.
- Watching an instructor solve a math problem does not mean you can solve one.
- Watching a nurse administer an IV does not mean you can do it.
- Reading a well-written novel does not make you a writer.

In each case, the professional — athlete, instructor, nurse, writer — completes the task with relative ease. These people have practiced these skills many, many times. They have developed extensive neural networks that activate when required. By watching their actions, you have started neural network development — the first step in the learning process. Practice is critical for you to become proficient at any task.

To illustrate a growing neural network, consider the progression of learning and applying mathematics.

Learning numbers and to count
→ learning addition and subtraction
→ learning multiplication and division
→ learning powers, exponents, and logarithms
→ learning that variables can represent numbers
→ learning geometry, trigonometry, and algebra
→ learning calculus
→ learning pure mathematics

By grade 10, you have learned about mathematical formulas. In subsequent grades, you manipulate and use increasingly complex formulas in your mathematics and science classes. These skills are required when learning calculus and pure mathematics.

Facilitating effective learning

Keep these observations in mind as you develop documents and presentations to ensure that the most information is conveyed.

1. For learning to take place, there must be an overlap between a learner's existing knowledge (existing neural network) and the new information. The new information builds upon and expands the neural network.
2. Information that is not linked to an existing neural network is quickly forgotten.
3. Repeated use is required to retain information. Repeated use activates and strengthens the neural pathway. Without repeated use, the neural pathway degenerates. From an educational perspective, you must practice to become capable and competent.
4. Learning is sometimes difficult. We may encounter a concept or skill that is difficult to master. When this occurs, the key to overcoming the barrier is to approach the problem from multiple perspectives until the concept or skill is learned and mastered. Barriers are encountered in all areas: sports, academics, business, …
5. Teaching, including peer instruction, forces a person to apply their knowledge and presentation skills simultaneously, creating links between them. This advanced learning is valuable in situations where quick recall of information is important: on exams, in future courses, in interviews, and in careers.

Effective teaching

Teaching and learning are complementary.

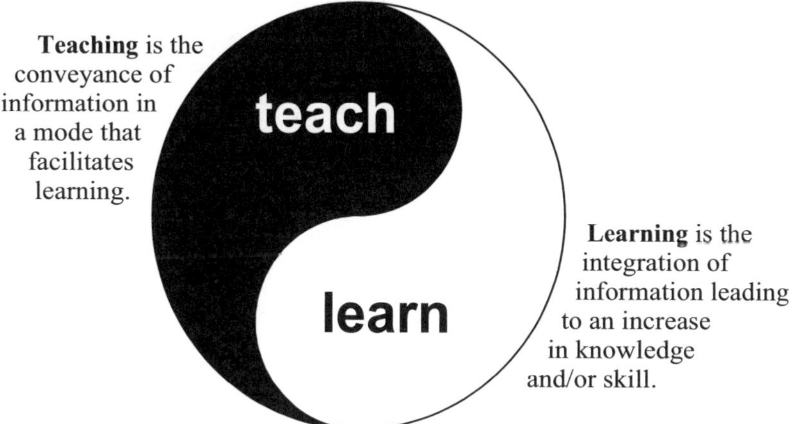

Figure 3.4 The complementary relationship between teaching and learning.

Instructors typically use instructional strategies that mirror their preferred learning style(s): we teach the way we learn. Students find learning easier when

- the instructor and student have similar learning styles
- the information being taught overlaps with the student's existing knowledge
- the student is interested in the information

When any of the above are not present, learning may be more difficult.* The student can still learn, but will feel somewhat frustrated. These are excellent opportunities for students to develop other learning styles. For instructors, teaching in an alternate learning style improves their ability to teach and learn in that style.

A significant problem occurs when a person learns incorrect information. The more layers of information built on this incorrect information — the more neural networks that link to the incorrect information — the harder it is to correct the erroneous information. From an educational perspective, in order to achieve the greatest understanding, students must be taught truthful information in a manner that allows future learning to build upon that information. For example, some high school chemistry instructors teach students that there are "seven different types of chemistry problems" and then teach ways of identifying the type of problem and possible solution strategies. While this strategy may work with the simple problems students face in high school, it introduces a misconception that science is strictly algorithmic, a misconception about chemistry, and it limits the students' ability to apply chemistry in a post-secondary environment and the real world. Students who need chemistry in their future studies and/or future careers will discover that chemistry problems are varied and complex. They must unlearn the strategy taught in high school and learn a more generalized problem-solving strategy. It would be pedagogically better to teach general problem-solving skills that students can employ in all their classes, from chemistry to home economics to auto mechanics.

* This is why one person may struggle with learning from a particular instructor, while other students, with different preferred learning styles, learn well from that instructor.

Summary

Learning is the formation of new connections between neurons, expanding the neural network. These connections are not permanent. Repeated use, in varied forms, is critical to strengthening the connections and expanding them to other networks. Without repetition and continued use, neural networks degenerate and knowledge is lost.

There are four learning stages — active testing, concrete experience, reflective observation, abstract hypothesis — that we cycle through when learning. The more we cycle through these steps, the greater our understanding of what is being learned, from riding a bicycle to complex algebra. While all four stages are critical to learning, a person often finds one stage easier than the others. This identifies a person's preferred learning style. Practice develops neural networks, improving our understanding of information, our ability to learn via other learning styles, and our ability to learn in general.

We are most comfortable teaching the way we best learn. When teaching, it is critical to not introduce incorrect and/or limiting information. Instead, we should teach in a way that builds on existing knowledge and allows future learning to build on the information being conveyed.

Additional resources ...

Zull J. The art of changing the brain. Sterling (VA): Stylus Publishing; 2002.

Kolb D. Learning style inventory. Boston: McBer and Company; 1985.

Persons wishing to determine their preferred learning style(s) can find several learning styles assessments online.

Chapter 4. Research methodology

Formally, there is a difference between *research* and *investigation.*

Research explores the unknown, with the results expanding the body of human knowledge.

Investigation explores existing knowledge, with the results consolidating and summarizing that knowledge.

Examples:

- A person preparing an essay on ozone depletion learns the relevant chemistry and atmospheric physics. This is an investigative project.
- A person joining a research group prepares a review summarizing the research in the field and uses this to decide on their research project. Their review is an investigative project.
- A person reviewing traffic accident data discovers a correlation between the accident rate and a change to the traffic light sequencing. This is a research project.
- An engineer looks up the properties of materials to select the best material for a new product. This is an investigative project.

Whatever project you engage in, your final work* must have both *style* and *substance. Style* is the language and formatting of the work. The primary goal of any scientific work is to convey information in a clear, coherent, concise, and precise manner. *Substance* is the information (content) you wish to convey. Without substance, you are preparing a work about nothing. Readers and audiences quickly realize this and are not impressed that someone is wasting their time.

All documents and presentations must have style and substance. Chapters 1 and 2 and 5 explore how to present that information, which provides the style. This chapter explores how to conduct a research/investigative project, which provides the substance.

Engaging in a research/investigative project

Learning how to conduct a research/investigative project is itself learned, as illustrate in Figure 4.1. Beginning in grade school, there is a gradual progression from an instructor-led to a learner-led environment. Graduate school is designed to train students to be independent learners and to develop independent research ideas and manage their own projects.

* *Work* is a general term that includes written documents and presentations. See Section 2.3 for details.

	Instructor led	⟶ Guided inquiry	⟶	Learner led
Selecting a research question	Learner uses question posed by instructor.	Learner refines question posed by instructor.	Learner poses question based on information from instructor.	Learner poses question.
Procedure development	Learner given procedure.	Learner given framework for procedure.	Learner told what data is needed and develops procedure.	Learner identifies what data is needed and develops procedure.
Data collection	Learner is aided with data collection.	Learner sets up equipment and collects data with supervision.	Learner sets up equipment and collects data with minimal supervision.	Learner sets up equipment and collects data.
Data analysis	Learner given steps to analyze data.	Learner given framework for analyzing data.	Learner told what information is required and formulates strategy.	Learner formulates analysis strategy.
Data interpretation	Learner told connections to scientific knowledge.	Learner given possible links to scientific knowledge.	Learner directed toward areas and sources of scientific knowledge.	Learner independently identifies links to scientific knowledge.
Reporting	Learner answers questions posed by instructor.	Learner answers questions in report format.	Learner informed what areas to cover in report.	Learner independently prepares report.

grade school → *undergraduate education* → *graduate school*

Adapted from *Inquiry and the National Science Education Standards*, Table 2.6, National Academy Press, **2000**.

Figure 4.1 The progression from an instructor-led to learner-led environment when conducting projects through a student's educational career.

Research/investigation development

Figure 4.2 illustrates the document development process.

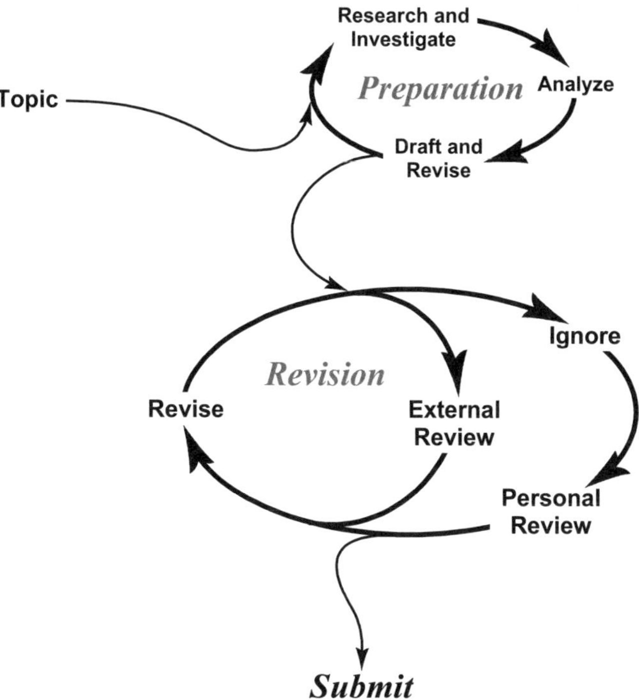

Figure 4.2 Document development begins with a topic. The preparation stage involves numerous development cycles before progressing to the revision stage, which again involves numerous cycles to prepare a quality document/presentation.

Time management is critical. You must have sufficient time to complete the project. As a rough guide, expect to spend

- 10 % of your time reviewing the literature
- 70 % of your time engaged in research or investigation
- 20 % of your time preparing documents and/or presentations

However, these do not occur sequentially. *Writing begins with the literature survey and continues through the entire project.* (Writing does not start in the last days before, or the night before, the document/presentation is due.) As your project progresses, you often need to consult additional literature resources and incorporate this new information into your project plan and into your developing documents. Reviewing your developing document reminds you of the overall project, improves your understanding of the background, and focuses your activities on the tasks required to complete the project.

Preparation involves reviewing the literature, developing a research method (for research projects), collecting and analyzing data (for research projects), and drafting document(s). Begin drafting your document(s) immediately by organizing your notes into paragraphs and sections. As you collect data and/or read additional resources, review and revise your developing document(s).

If you are engaged in a research project, you may wish to start with the *Introduction* and *Procedure* as you investigate the history and develop an experimental method for your research. If you are engaged in an investigative project, you may wish to start with the body chapters as you investigate the topic from different perspectives.

Once you have completed the project and written the first draft, read the document in its entirety and revise it to ensure there is a logical and coherent train of thought. Do not be surprised if your first revision is a substantial rewrite. The goals of this self-review are to

- remove redundant and extraneous information
- move information to more appropriate locations
- clarify confusing sentences and paragraphs
- adjust the language and terminology to the appropriate level

Strategies for self-review include:

- Read your work aloud. This allows you to simultaneously read and hear the logical progression in your prose. Recall that you should be writing for the reader. Your written work should sound professional, as if you are explaining your work to a person who is interested in it. Depending on the document you are preparing, the person could be
    - a colleague involved in similar research (laboratory report, research proposal, scholarly article)
    - a colleague within your discipline (essay, thesis, poster, oral presentation)
    - a person interested in your work (project report, document for public audiences, oral presentation)

  and you must adjust your language accordingly.
- If you find yourself referring to other parts of the document, revise the section to bring the relevant information to the required location.
- If you find yourself rereading a sentence/paragraph to understand it, it must be revised. Common causes include overly complex sentences, poor grammar, misplaced modifiers, poor linkage of old and new information, and paragraphs that contain more than one concept.
- Ensure that every sentence and every paragraph conveys information, that excess words are removed, and that repetition is minimized.

Put simply, if you do not understand what you have written, your reader does not have a chance. They will get frustrated and give up reading your work. Recall that you are striving for a clear, coherent, concise, and precise document.

Wait a few hours to a few days (depending on the length of the document) before repeating your review. This delay makes you more objective. You may be surprised that what made sense when preparing the work does not make sense when you read it again. Repeat your self-review until you believe there is a logical and coherent train of thought in the work. Once *you believe* you have a reasonable work, external revision begins.

Revision involves giving your beloved work to colleagues and asking them to pick it apart. Their job is to identify grammatical errors, information that is redundant and irrelevant, and sections that do not flow. You should put your work aside until you receive the external feedback, then review it again yourself.

If a work requires significant revision, the reviewer may substantially reformat sections and/or write large amounts of material for inclusion in the work. The acceptability of this depends on the environment.

- In student academic environments, where a single student is expected to prepare the work, incorporating someone else's contributions may constitute academic misconduct or plagiarism. Using large passages written by the reviewer is not permitted. It may be acceptable if the student acknowledges the reviewers and their contribution to the work documented. Correcting spelling and grammar errors and rewriting sentences and paragraphs that a reviewer identified as not making sense is likely acceptable. Ultimately, what is acceptable (or not) is at the instructor's discretion.
- In professional academic environments, persons who contribute substantially should either be listed as an author or in the acknowledgements section, depending on the nature and extent of their contributions.
- In employment environments, it is sometimes acceptable for many people to contribute to a document and not be credited at all.

The difference between the expectations of students to independently prepare material and the collaborative professional environments may seem extreme until you consider the meta-learning expectations. Students must know how to prepare documents independently so that they can contribute to the team documents. Preparing scientific documents develops your ability to identify and evaluate relevant works, to synthesize new knowledge from research and place it in the context of existing knowledge, and to communicate this knowledge to others.

The nature of your project dictates how long you spend preparing and revising it. At least one other person should review a laboratory report. A research article will be reviewed by your colleagues (more than once), by your supervisor (more than once), and by two to five anonymous reviewers chosen by the journal.

> As a graduate student, I was asked to watch a fellow student's rehearsal for their defense. They gave a good presentation, but I commented that their presentation contained too many images of the research setting (mountain ranges and pollution source) and not enough on the science (methods, analysis, and results). The student and their friends instantly spurned me for "criticizing" the presentation. However, the lack of science in the presentation also concerned the defense committee who required the student expand these sections before approving the thesis.
>
> The moral of this story is that, if you ask someone to review your work, they have given you their time, and you should consider their comments rationally. (Chapter 6 provides more information on the review process.)

Submission only happens once: you submit your document to the world. The "world" could be your laboratory instructor, your instructor and classmates, the committee who reviews your undergraduate research report, the conference attendees who attend your presentation, or the readers of a scientific journal. You have one chance to submit quality work.

4.1 Research methods

Since before recorded time, humans have been trying to understand natural phenomena. These endeavors have developed into the realm of *science*: the systematic study of the world around us. *Scientists* engage in *research* to explore and understand what is currently unknown, and their results and understanding add to the ever-expanding body of scientific knowledge (see Figure 1.4 on page 9).

The above paragraph introduces several important concepts:

- Research is an exploration of the unknown that leads to newfound understanding.
- Research is conducted in a systematic and reproducible manner so that others can repeat and verify it.
- Research results are communicated so the newfound understanding increases global scientific knowledge.

Scientific results are unchanging facts, but the interpretation of those results may vary from person to person and over time. With additional information, often obtained from additional research, the interpretation of specific results may change. In this manner, the body of scientific knowledge is ever-evolving and self-correcting.

Research methods can be divided into five broad categories. Scientists use these methods separately and in combination to understand the phenomena they are investigating.

Experimentation: the relationship between two or more variables is determined by controlling one or more variables, measuring the response of the other variable, and holding all other parameters constant.

> Chemists prepare calibration curves to determine sample concentrations.
> Physicists identify particles produced in high-energy collisions.

Comparison: the relationship between two or more variables is determined by exposing samples to different variables and measuring the response.

> Drug companies test new drugs against a placebo.
> Biological experiments have a control and one or more test groups.

Observation: phenomena are observed without interference from the scientist.

> Darwin observed variability in finch species on the Galapagos Islands.
> Copernicus observed the movement of planets, which led him to propose that the sun is the center of the solar system.

Modeling: physical or computer-based models are used to mimic natural systems.

Organic chemists make heme analogues to study the iron:oxygen interaction.
Biochemists use computers to model protein receptor sites in order to understand the binding process and identify molecular entities that may bind to the receptor.
Environmental scientists model ocean currents to understand nutrient flow.

Theoretical: theoretical models are developed to explain the results of other types of research.

Mathematicians develop mathematical functions that are then used by scientists.
Physicists are developing a *theory of everything* to unite the known forces.
Chemists develop theories of chemical bonding and intermolecular interactions.

Scientific method

There is no single procedure for conducting research — no single *scientific method.* Rather, research begins and proceeds via the general framework in Figure 4.3. In reality, there is no beginning or end to research: research builds on work done previously, and publishing the results allows others to conduct further research.

If I have seen further, it is by standing on the shoulders of giants. — Isaac Newton

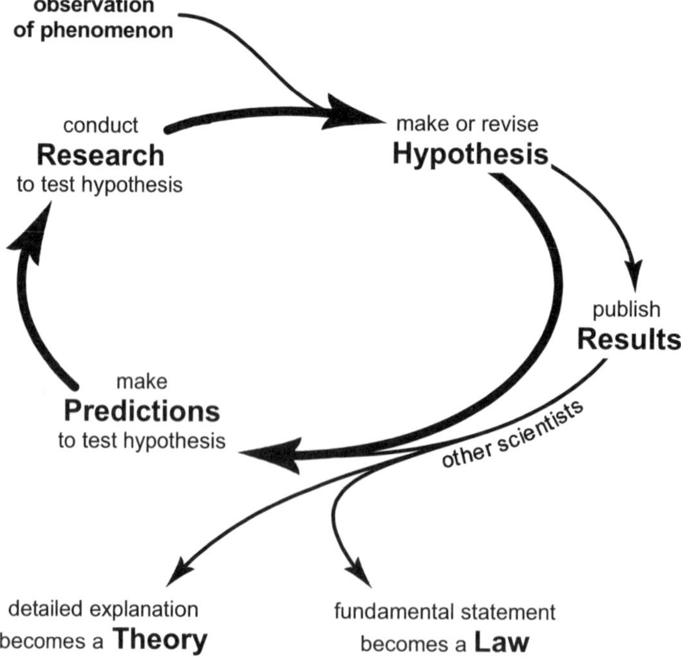

Figure 4.3 The scientific method: a framework for conducting research.

We make *observations* every day — the traffic light is going to turn red, what a beautiful sunset, it will probably rain tomorrow — but these observations do not advance science until one looks beyond the observation:

> How do drivers react when they suspect the light will turn red?
> What causes the colors in the sunset?
> What is it about the environment that makes me believe that rain is coming?

More generally, these questions are

> Why does ____ occur?
> What happens if ____?
> Can we make ____ happen?

All of these questions lead to the formation of one or more hypotheses.

A *hypothesis* is a proposed and testable explanation of phenomena. A good hypothesis explains the phenomena and predicts the outcome of future research. If the predictions are incorrect, the hypothesis is rejected.* If the predictions are correct, the hypothesis gains strength and credibility. The predict–test–update hypothesis cycle continues to refine the hypothesis to more accurately explain the phenomena and to establish the applicable range of the hypothesis. Once understanding is achieved, the results are published. Publishing adds this information to the body of scientific knowledge for other scientists to evaluate, support, criticize, and build upon. Some hypotheses are given below.

> Assuming that drivers want to be safe on the road, drivers will slow down in anticipation of the light turning red.
> The colors in the sunset occur because the Earth's atmosphere is acting as a prism.
> Together, increasingly thick and dark clouds, falling temperatures, and high humidity suggest that precipitation is forthcoming.

A *theory* is a set of rigorously tested statements or principles that explain phenomena and can be used to make predictions about phenomena. Evolving from a hypothesis, a theory is accepted by the scientific community as a correct explanation of phenomena, but it can still be disproved. The transition of a hypothesis into a theory may take millennia. Established theories include

- the theory of evolution
- quantum theory
- the theory of relativity

* Proving a hypothesis wrong has led to significant advancements in science: phlogiston (1667), aether (1704), lipid hypothesis (1913), cold fusion (1989), …

A scientific *law* is a concise statement that describes the relationships among phenomena. Laws are often mathematical. Accepted laws include

- the ideal gas law
- the law of conservation of energy
- the law of conservation of mass
- the law of universal gravitational

More succinctly: a *law* explains *what* occurs and a *theory* explains *why* something occurs.

Steps to conducting a research project

Formulate the research question: the research question clearly defines the objectives of the research: given an observation, what are you trying to understand? The question must be specific and focused. While conducting the literature survey and developing a research method, you may discover that your question is already answered, too broad given the available time, and/or that the necessary resources are not available. In all these cases, the research question must be modified.

Literature survey: learning what is already known about the research question. Your question may have already been answered or partially answered. By determining what is already known, you ensure your research explores the unknown.

Hypothesis development: with an understanding of what is known in this field, you propose an explanation — a hypothesis — for the phenomena.

Research method design: methods are proposed and developed to test the hypothesis. These methods are often variants of those found in the literature survey. Before beginning full-scale research, a pilot study is conducted to ensure the method yields the desired data. Refinements to the method often improves the quality of the data. The final method is documented in sufficient detail so that others scientists are able to reproduce the research.

Data collection: data is collected. Observations are recorded to understand the data and aid in the analysis.

Data analysis: the data is analyzed and conclusions are made regarding the correctness of the hypothesis. Additionally, an objective meta-analysis of the research method design, data collection, and data analysis is conducted. You *do not* want to publish erroneous results and conclusions.

One cycle of the scientific method in figure 4.3

Very often, the data analysis identifies something interesting that may affect your hypothesis. You then repeat the above steps until you have a comprehensive understanding of your observations and how they integrate into the broader scientific field.

Reporting: the results are reported in scholarly articles, at scientific conferences, in theses, and to the public.

Steps to conducting an investigative project

An investigative project has many of the same steps as a research project, but the focus of an investigative project is on the literature survey and analysis thereof.

Formulate the investigative question: the investigative question clearly defines the objectives of the investigation: what are you trying to understand? The question must be specific and focused. While conducting the literature survey, you may propose a question that you discover is already answered or too broad given the available time. In these cases, the investigative question must be modified.

Literature survey: learning what is known about the scientific field. The literature survey is the major component of an investigative project. You are expected to identify all the significant contributions to the field.

Literature analysis: you analyze the information to understand how this field integrates into the related areas of science, understand the theory underlying the field, understand the research methods used by the research groups, determine what is commonly agreed upon, identify poorly understood phenomena, and compare and contrast the hypotheses proposed to explain the poorly understood phenomena.

Reporting: the results are conveyed in reports, essays, presentations, and to the public.

Literature survey

It is important to determine what is already known about your topic so that you can build on the work done by others. *Scholarly articles* (often called *papers* or *primary literature*) are the primary means of publishing new scientific information. *Scholarly books*, such as textbooks and theses, often present greater explanations of complex theories than is reported in scholarly articles. However, scholarly books typically undergo less peer review than scholarly articles.

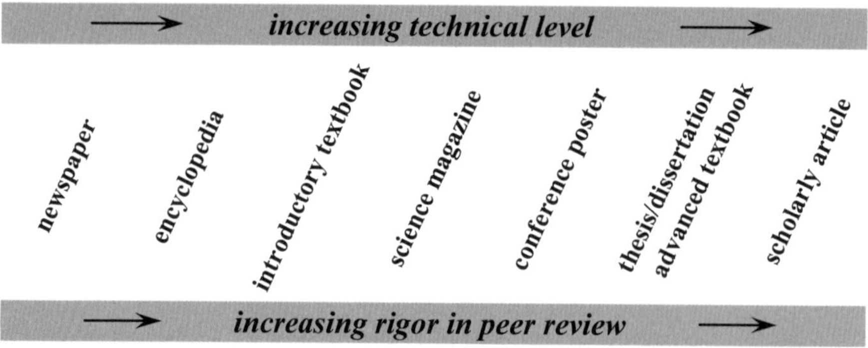

Figure 4.4 Common publications that have some degree of peer review.

When conducting a literature survey, you should endeavor to use scholarly articles and scholarly books. Scholarly articles provide the best impartial knowledge on your topic. One type of scholarly article is a *review article*, which reviews and summarizes the research and current understanding of a scientific field. A recent review article provides an excellent overview of the field and provides references to more focused scholarly articles that may be relevant to your topic.

> *It is that which we do know which is the greatest hindrance to our learning, not that which we do not know.* — Claude Bernard

There are many resources available for you to conduct a literature survey. A common way to begin is to conduct a literature search of one or more databases.

- University libraries have access to databases that index more rigorous publications (conference proceedings, scholarly books, and scholarly articles).
- Schools and public libraries have access to databases that index less rigorous publications (newspapers, science magazines).
- Internet search engines often link to scholarly articles.

The resources available to you vary depending on your institution. The best advice is to *talk to the librarian*, who can direct you to the most appropriate resources for your project.

Scientific databases index the titles and abstracts of articles from specific disciplines (science, medicine, social science, etc.). Common scientific databases available at colleges and universities include *SciFinder*, *ScienceDirect*, *MedLine*, and *Web of Science*. *Google Scholar* is a free database available at http://scholar.google.com.

Once you find an article, most databases link to both the articles cited and the articles that cite the article you found. This allows you to identify the articles that introduce a topic and identify the articles that present the latest understanding of that topic.

Effective use of search engines

All search engines have numerous advanced abilities:

- searching for specific phrases
- including and excluding words and phrases
- including wildcards in words

The syntax for conducting advanced searches requires an understanding of Boolean logic. This format is used by most search engines, including Google, Bing, and most libraries. Boolean logic has the following features:

- specific phrases must be placed within quotation marks
- all terms and phrases are required
- to exclude a word, precede it with a negative sign
- the asterisk wildcard refers to any character or characters

The Help section of these search engines presents additional features to improve your search.

Internet search engines link to scholarly books and articles, but also link to personal and commercial websites. You should always be wary of bias when reading personal and commercial websites. Bias is often encountered when investigating controversial topics (abortion, environment, GMOs, religion, politics, …) and commercial topics (fitness, health, technology, …).

A word on Wikipedia: in general, Wikipedia is a good readily accessible resource. But because anyone can create and edit an article, Wikipedia is subject to bias and misinformation. Controversial topics side with popular beliefs at the expense of factual information.* For these reasons,

* The reason Wikipedia sides with popular belief is because Wikipedia's verifiability policy: if the majority of sources present incorrect information, Wikipedia will present incorrect information. For example, a 17th century Wikipedia would present a geocentric (Earth-centered) model of the solar system and aggressively exclude the heliocentric (sun-centered) model. For a recent example, see

Messer-Kruse T. The 'Undue Weight' of Truth on Wikipedia. The Chronicle of Higher Education [internet] [cited 08 November 2012]. Available from: http://chronicle.com/article/The-Undue-Weight-of-Truth-on/130704/

Wikipedia articles should be used with caution and the information confirmed in scholarly articles. The references in Wikipedia articles often link to scholarly articles. These links can also be used to judge the neutrality of the Wikipedia article itself.

Reading academic works

Reading academic works is different from reading literary works. In academic works, you are reading for understanding. Ideally, the author has written clearly and concisely, putting meaning into every word and sentence. You must understand every word to understand the sentence, paragraph, section, and work. To gain this understanding requires reading the work multiple times and consulting other resources (dictionaries, textbooks, reference books, and colleagues) to understand what the author has written. The author will cite other works, so if you do not understand a concept or if a concept is especially important to your project, it will be valuable to read these work as well.

When reading academic works, it is important to obtain a general understanding of the entire academic work, but you will probably be interested in only one or a few aspects of the research. For most readers, the *Introduction*, *Discussion*, and *Conclusion* sections are important because these sections explain the current understanding of the scientific field. For readers wishing to conduct similar research, the *Method* and *Results* section are also important.

Article summaries

As you read academic works, it is valuable to prepare summaries that extract and condense the important information. The summaries are short documents that you can use to quickly refresh your memory of the full article. Each summary should include the full citation of the article, an overview of the work conducted, and detailed notes on the information in the article, especially the information relevant to your project. Section 5.4 contains additional information on summarizing scholarly articles.

Consider using citation management software to electronically organize your resources and summaries. Common citation managers include ProCite, RefWorks, and EndNote. These programs

- store the full citation of the resource
- store an electronic copy or link to the resource
- store your summary
- are searchable

- integrate with online databases to obtain additional scholarly articles as required
- integrate with word-processing software to automatically generate the *Bibliography* or *References* or *Works cited* section of your work in the required style

Reporting

When reporting on your work, your *Introduction* must present the relevant background information that places your project in context of what is already known in the scientific field. The *Method and materials* section must be sufficiently complete for an equally skilled person to reproduce your experiments to confirm your results or build on your work.

Readers often focus on your *Results* and *Discussion* sections. These sections must present and analyze the data in an unbiased and logical manner. You must convince the reader that your data is correct and your analysis and interpretation are reasonable given the available data. This is especially important when reporting on new, novel, and/or controversial topics.* Readers — scientists, politicians, corporations, and the public — may question your data, your analysis, and your neutrality. Unfortunately, not everyone is open-minded and accepting of alternative perspectives. Some may want to discredit your research/investigation, therefore your discussion and conclusions must be supported by your data and convincingly argued. Furthermore, it behooves you to acknowledge limitations in your data and acknowledge competing interpretations, and then explain why your interpretation is most appropriate.

Contrary to the term, research/investigative projects cannot *conclude* or *prove* anything. (In mathematics, however, it is possible to prove a relationship: a mathematical proof.) Experimental results may support the hypothesis, but there may be other interpretations that are equally valid. Indeed, science continually develops better models that explain both the historical data and the anomalies that early models could not explain.

* Students commonly investigate controversial topics: pollution, climate change, alternative energies, drug legalization, food additives, and many others.

Peripheral information

As you engage in a literature survey, you will learn about related topics. This peripheral knowledge adds to your understanding of the field and how it affects society, and may be valuable when you answer questions on your work. However, peripheral knowledge is not included in your *focused* scientific work. First drafts commonly contain irrelevant and superfluous information that must be removed through the review process. However, what is considered relevant to a given project depends on the author, reviewers, and readers.

A word on school laboratory experiments

As you proceed through high school and university, you will attend laboratory classes where you start and complete an experiment in one to four hours. Within a week, you will analyze the data and submit a report. You may lose marks if the experiment does not work or if your results are not close to the “correct” value.

This is not research!

First, research is an investigation of the unknown: there are no correct answers to compare your results with. Second, research takes weeks to months to years, depending on the project. Third, failure is common and, while frustrating, not a cause for punishment. Fourth, experiments are repeated multiple times to optimize the procedure, collect additional data, and better understand the phenomena.

Laboratory classes are valuable because they introduce the methods used in that discipline to conduct research and expose students to scientific equipment, the laboratory environment, data collection in laboratory notebooks, data analysis, and preparing laboratory reports. Laboratory courses provide students with the technical knowledge and skill to safely perform accurate and precise measurements. As you proceed through university science courses, the laboratory experiments become increasingly like “real” research in that you may be required to conduct a literature survey and develop a procedure, and the experiment may take several laboratory periods to complete. In some experiments, you may analyze systems for which the answer is not known.

The entire laboratory program — from high school to fourth-year science courses — prepares you to engage in research.

4.2 Research ethics

Scientists must endeavor to establish and maintain a research philosophy that avoids all forms of scientific misconduct:

- unethical treatment of subjects
- fabrication/falsification of data
- exclusion of data
- plagiarism
- fraud
- bias

Scientific misconduct concerns all scientists because it draws into question the impartiality of all scientists and all published scientific work. Scientists should continually monitor themselves and their colleagues for possible lapses in conduct. If misconduct is identified early, corrective action can be taken and damage minimized. Proven allegations of misconduct have cost scientists their careers, embarrassed scientific journals, and initiated lawsuits against scientists and their institutions.

Scientists engage in unethical practices for many reasons:

- Research funding is getting more competitive, so there is increased pressure to embellish results to get funding.
- Funding sources with a stake in the results of the research, such as pharmaceutical companies, threaten to withdraw funding if the companies' preferred results are not obtained.
- The pressure to "publish or perish" causes scientists to conduct research and publish work faster, with fewer checks on the work.
- Research assistants may not fully comprehend the research process or the ramifications of violating research ethics.
- Scientists desiring prestige and recognition may embellish results.

Scientific misconduct may be discovered during the peer review process. Since the reviewers are also experts in the scientific discipline, they are aware of what is reasonably possible. They will diligently scrutinize extravagant claims. After publication, the author's words are a permanent record of what they did. Other researchers may endeavor to repeat the research, and become concerned if they obtain different results. Additionally, automated search and compare tools are becoming increasingly sophisticated. Copying something from an obscure book today may be caught in a few years as technology improves.

Anomalous data

Occasionally, when collecting data, you may have one or more data points that do not follow the apparent trend. You cannot arbitrarily choose to exclude data! You can only exclude data under two conditions:

1. something happened to the sample that could have affected the results (spilled, contaminated, etc.) and this is documented in your laboratory notebook

   Laboratory notebook: sample 8 may have been contaminated when <chemical> accidentally spilled and splashed into my work area.

   Laboratory notebook: data collected after 10:20 PM may have a systematic error because the detector was bumped and had to be realigned.

2. a data point fails a statistical retention test (see page 108).

If data points do not follow the apparent trend, it is possible that there were uncontrolled variables in the experimental method or that something else is occurring. Exploring anomalous experimental results may also lead to new discoveries.

Code of research ethics*

Honesty: strive for honesty in all scientific communications. Honestly report data, results, methods and procedures, and publication status. Do not fabricate, falsify, or misrepresent data. Do not deceive colleagues, granting agencies, or the public.

Objectivity: strive to avoid bias in experimental design, data analysis, data interpretation, peer review, personnel decisions, grant writing, expert testimony, and other aspects of research where objectivity is expected or required. Disclose personal or financial interests that may affect research.

Integrity: keep your promises and agreements, act with sincerity, and strive for consistency of thought and action.

Carefulness: avoid careless errors and negligence; carefully and diligently examine your own work and the work of your peers. Keep good records of research activities, such as data collection, research design, and correspondence with agencies or journals.

* Adapted from

Resnik, DB. What is Ethics in Research & Why is it Important? [internet]. NIEHS-NIH; 2011 [cited 22 March 2012]. Available from http://ww.niehs.nih.gov/research/resources/bioethics/whatis/ Used with permission.

Openness: share data, results, ideas, tools, and resources. Be open to criticism and new ideas.

Respect for intellectual property: honor patents, copyrights, and other forms of intellectual property. Do not use unpublished data, methods, or results without permission. Give credit where credit is due. Give proper acknowledgement or credit for all contributions to research. Never plagiarize.

Confidentiality: protect confidential communications, such as papers or grants submitted for publication, personnel records, trade or military secrets, and patient records.

Responsible publication: publish in order to advance research and scholarship, not just to advance your own career. Avoid wasteful and duplicative publication.

Responsible mentoring: help to educate, mentor, and advise students. Promote their welfare and allow them to make their own decisions.

Respect for colleagues: respect your colleagues and treat them fairly.

Social responsibility: strive to promote social good and prevent or mitigate social harms through research, public education, and advocacy.

Non-discrimination: avoid discrimination against colleagues or students on the basis of sex, race, ethnicity, or other factors that are not related to their scientific competence and integrity.

Competence: maintain and improve your own professional competence and expertise through lifelong education and learning; take steps to promote competence in science as a whole.

Legality: know and obey relevant laws and institutional and governmental policies.

Animal care: show proper respect and care for animals when using them in research. Do not conduct unnecessary or poorly designed animal experiments.

Human subjects' protection: when conducting research on human subjects, minimize harms and risks and maximize benefits; respect human dignity, privacy, and autonomy; take special precautions with vulnerable populations; and strive to distribute the benefits and burdens of research fairly.

Before conducting research involving humans or animals, you will need approval from an Institutional Review Board (also called the Research Ethics Board, Ethical Review Board, or Independent Ethics Committee). The Board provides an independent assessment of proposed research to ensure the research minimizes harm to human subjects and/or humanely treats animals. Boards have the power to approve, modify, or reject research proposals.

Depending on your project, you and your supervisor may need to apply to the Board as you design your research project.

Chapter 5. Documents and presentations

In your career, you will do many of the following:

- document the work you are doing
- discuss your work with colleagues and collaborators
- write memos to your supervisor and to the people you supervise
- apply for project approval and/or funding
- present your work to coworkers and at conferences
- prepare instructions for others to follow
- prepare reports for clients
- prepare articles for journals and magazines
- write press releases

In all cases, clarity, coherence, concision, and precision are of the utmost importance. You must communicate information in a manner the reader understands and expects. Figure 5.1 illustrates the usual audience of different public and scientific documents.

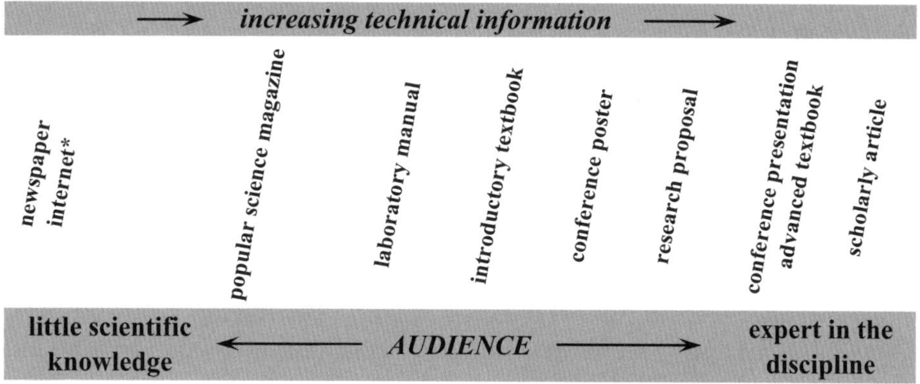

Figure 5.1 An overview of the level of information in and the usual readers of common documents.

A common misconception is that there is only one correct way to write something. In fact, there are *many* correct ways and many incorrect ways. Companies, scientific journals, newspapers, and instructors adopt guidelines for consistency, and you must abide by those guidelines when preparing documents for these audiences. However, no single guideline is substantially better than any other.

* The internet does contain highly technical information, but it also contains well-presented misinformation. Because documents on the internet are not reviewed, the credibility of information on the internet should always be questioned.

Figure 5.2 illustrates the similarities between an academic work and a literary work.

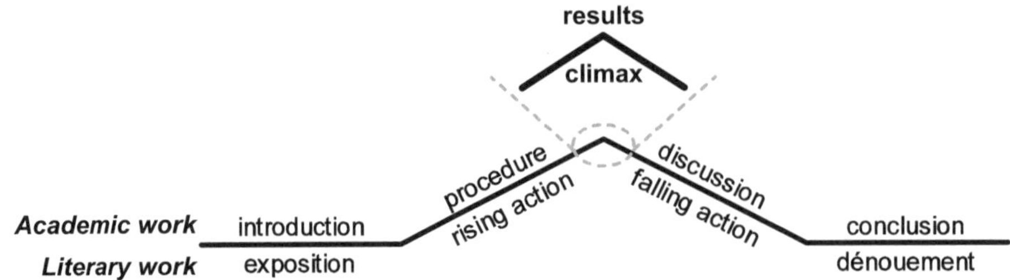

Figure 5.2 The similarity in structure of a scientific document and dramatic story.*

Literary works and academic works have several similarities:

- Both tell a coherent story.
- The introduction foreshadows the results and conclusions.
- There should not be any irrelevant information.

For example, consider the television show *CSI* (*Crime Scene Investigation*), which bridges the literary and academic realms. The incriminating evidence is presented clearly and logically so that the viewer can understand how the conclusions were reached. You must do this with your data.

Document formatting: general guidelines

Most scholarly documents have minimal formatting, and the following configuration will work for them. Once configured, save it as a document template and use it for all the scholarly documents you prepare. (Appendix B.1 provides instructions for configuring your page layout, text styles, and creating templates.)

Page layout

- 2.5 cm margins on all sides
- 1.0 cm gutter margin if the document is stapled or bound
- If your document is double sided, mirror the margins
- *header*: 2.0 cm from edge; *footer*: 1.8 cm from edge

* Adapted from

Hudson R. Exploring and teaching concision in scientific writing, Canadian Society for Chemistry conference; Calgary (AB). 26–30 May 2012. Used with permission.

Text styles

- Use a minimum number of fonts: one font for the title and headings, another font (or the same font) for the text.
- 1.0 or 1.5 spacing (1.0 is single spaced)
- *headings*
    - maximum of two heading levels for most documents
    - a single font for all headings levels (Arial or Times New Roman are common)
    - adjust the font size, font formatting, and space before/after to create white space between sections
- *lists*, *equations*, *examples*: all are indented from the surrounding text
- references are formatted in the style required by the instructor, employer, or publisher

The following text styles will give your documents a simple, logical layout with constant formatting throughout the document. Both of these make your document easier to read.

- *Title*: 16 point; bold; space after: 18
- *Heading 1*: 14 point; bold; space before: 18; space after: 6
    - if starting a page: space before: 0
- *Heading 2*: 12 point; bold; space before: 12; space after: 3

} Arial or Roman font (select one, use for all)

- *Normal text*: 12 point; Times New Roman font; spacing: at least 16; space before: 3; space after: 3
- *Lists, equations, examples*: <same as Normal text>; left indent: 1.0 cm; space before: 0

The Spacing:Before and Spacing:After settings create ***white space*** between the objects (text, figures, tables) on the page. White space is important because it identifies what objects go together conceptually. Too little white space makes the page cluttered and difficult to read.

For example, consider headings: the large space before the heading creates a visual break between the sections, and the small space after the heading links the heading to the following text.

Document improvement: spelling and grammar analysis

Word processors can analyze your document for both spelling and grammar, and recommend changes to improve both. Repeat this analysis regularly as you prepare your document, but do not automatically accept the recommendations. While most recommendations are beneficial, the word processor is not programmed with all scientific terms and proper nouns, and sometimes interprets scientific phrases incorrectly. Critically review all recommendations to determine if the suggestion retains the same meaning and improves the clarity, cohesion, concision, and precision of your work.

Document improvement: document readability

Word processors can calculate the readability of a document. Numerous factors affect the readability, including average word length and average sentence length. The underlying premise of readability analyses is that shorter words and shorter sentences are easier to comprehend.

Word processors often calculate the Flesch-Kincaid readability statistics.[*] Two statistics are reported:

- *Reading ease*: the higher the reading ease, the easier the document is to comprehend.
- *Grade level*: the higher the grade level, the harder the document is to comprehend.

To interest and retain readers, the readability statistics should match the expectations of the reader. Too low a reading ease (harder) and the reader will become confused by the document. Too high a reading ease (easier) and the reader may feel belittled by the document or question the competence and knowledge of the author. Table 5.1 presents recommended readability statistics for different documents.

Table 5.1 Recommended readability statistics for scientific and public documents. The correlation between reading ease and grade level is approximate.

Document	Reading ease	Grade level
graduate documents; scholarly articles	< 30	> 14
undergraduate documents	30 – 40	12 – 14
popular science articles; technical reports	30 – 50	10 – 14
general audience (newspapers, websites)	50 – 70	6 – 10
children's books	> 70	< 6
Communicating Science	47	10

Section B.1 shows how to calculate the readability statistics for your document.

* Wikipedia. Flesch–Kincaid readability tests (internet). Accessed 12 July 2013. Available from http://en.wikipedia.org/wiki/Flesch–Kincaid_readability_tests

5.1 Notebooks

Many careers require the keeping of a notebook: science, journalism, law enforcement, etc. Notebooks are valuable for both research and investigative projects. The notebook is a repository of all the information you have collected on a project, recorded when you obtain the information. The laboratory notebook is a permanent record of what you did and what you observed, written while you conducted experiments. This allows you to accurately document materials, procedure, modifications, and observations, and to show your timeline for completing the work.

The information in your notebook forms the basis for every document and presentation you prepare on the project.

> The research notebook is a legal document in patent cases. The information in a notebook is critical to confirming when and how a discovery or an invention occurred. Errors in maintaining a notebook can seriously damage a discovery claim or patent application, and may void an existing patent.

Electronic laboratory notebooks are a recent technological development that is gaining acceptance in academic and industry. The software is designed to maintain a permanent record of what is recorded. They are convenient because they can collect data directly from instruments, are searchable, allow multiple scientists to record their data in the same location, and can automatically create an off-site copy.

Formatting a laboratory notebook

The following guidelines will help you prepare and maintain a quality laboratory notebook.*

* Adapted from

Protecting your intellectual property. University of Oklahoma: Office of Technology Development; [cited 11 July 2011]. Available from: http://otd.ou.edu/pdfs/Protecting_Your_Intellectual_Property.pdf Used with permission.

Project: *Using a laboratory notebook* 1 – 5
book page

continued from page: *start*

The notebook

- Use bound notebooks with numbered pages.
- Number notebook volumes sequentially and with the start and end-use dates.
- Use NCR (non-carbon recording) paper or have a policy of photocopying pages once they are complete. At pre-determined times (end of day, end of week, end of experiment), submit copies to the laboratory supervisor for review and off-site storage. In the event of an accident, having an off-site copy is critical to continuing from where you left off.
- Do not add or remove pages. Adding or removing pages destroys the credibility of all the information in the notebook.
- Keep lab notebooks in a safe place when not in use.
- Keep lab notebooks for at least 20 years after completion.

Notebook organization

- Reserve the first four pages for a table of contents. Add an entry to the table of contents every time you start a new experiment.
- Complete the notebook chronologically. Do not leave space to complete a project. If you are working on multiple projects, draw a dividing line each time you switch projects.

Legal considerations

- Write legibly! It must be possible for you and others to understand what you have written and to reproduce your work.
- Write information directly into your laboratory notebook. Copying information into your notebook later could introduce errors and you could lose the pages before the information is copied.
- Mistakes: make corrections by drawing a single line through the entry and initialing the correction. (Do not erase or obliterate mistakes.)
- Record significant events: successful testing of a new invention, synthesis of a new compound, the discovery of a new organism, ... Legally, this is "reduction to practice" and the date of such accomplishments is important. Photographs and test data are valuable to confirm these events.

continued on page: 6

Signature	Date	Witness	Date
Roy Jensen	20 June		

Project: *Using a laboratory notebook* 1 – 6
book page

continued from page: 5

Legal considerations (continued)

- *If something doesn't work, don't write dismissive statements in the notebook. Legally, it can be argued that you abandoned the idea.*
- *Use permanent ink that does not bleed. Most inks are stable in water, but many inks bleed in common laboratory solvents, as shown below.*

Pen	Abuse treatment						
	Control	Erasure	Water	Methanol	Ethanol	Acetone	Baked
Bic Accountant fine point (red)	123	123	123				123
Bic Accountant fine pt (black)	123	123	123				123
Bic Round Stic med (black)	123	123	123				123
Cross fountain pen (blue/black)	123	123	123	123	123	123	123
Dixon Ticonderoga 1388-2 soft pencil	123	123	123	123	123	123	123
Pentel Hybrid Gel Roller (black)	123	123	123	123	123	123	123
Pilot G-2 07 (black)	123	123	123	123	123	123	123
Sakura Gelly Roll fine (black)	123	123	123	123	123	123	123
Sakura Gelly Roll fine (blue)	123	123	123	123	123	123	123
Sakura Gelly Roll XPGB (blue)	123	123	123	123	123	123	123
Sakura Gelly Roll XPGB (green)	123	123	123	123	123	123	123
Sakura Gelly Roll XPGB (red)	123	123	123	123	123	123	123
Sakura Pigma Micron .45 mm (black)	123	123	123	123	123	123	123
Sanford Sharpie extra fine (black)	123	123	123	123	3		123
Sanford Sharpie extra fine point (red)	123	123	123				123
Sanford Sharpie ultra fine point (blue)	123	123	123				123
Sanford Uni-Ball Gel RT Med (black)	123	123	123	123	123	123	123
Sanford Uni-Ball Vision fine (black)	123	123	123	123	123	123	123
Sanford Uni-Ball Vision fine (blue)	123	123	123	123	123	123	123
Sanford Uni-Gel RT fine (blue)	123	123	123	123	123	123	123
Zebra Sarasa 0.7 (blue/black)	123	123	123	123	123	123	123

(Courtesy of Colin Purrington. Used with permission.)

- *If you do not work in the laboratory for a period of time (conferences, administrative work, vacation, ...), document the reasons and the dates spent away from the laboratory.*
- *When two or more investigators are working on the same project, they should enter all the data into one notebook, not separate notebooks. One notebook should contain the complete record of the project. Investigators should initial and date their own entries.*

continued on page: 7

Signature	Date	Witness	Date
Rory Jensen	20 June		

Project: Using a laboratory notebook 1 – 7

book page

continued from page: 6

Documenting projects

- When starting a project, give a brief overview of the project, including the relevant theory and rationale behind the project.
- Include a flow chart of the project and detailed flow charts of each experiment.
- List the reagents (including manufacturer and lot number), organisms (supplier, line history, delivery conditions, storage conditions), and equipment used (make and model number).
- List the experimental setup, plus any notes or references to why you selected this experimental setup.
- Detail the experimental procedure or reference a published procedure.
- Detail tweaks made to the procedure during the experiment.
- If something must be inserted, such as a picture, permanently affix it to the notebook and have its placement witnessed by another person.
- Date and sign every page and have your laboratory supervisor sign each page. The laboratory supervisor should not be directly involved in the project.

Information that should be recorded

- Record everything when it is measured or observed.
- Record experimental details: equipment settings, temperature, pressure, flow rates, light levels, environmental conditions, etc., and how they changed during the experiment.
- Record experimental data (it is often convenient to tabulate data).
- If data are collected electronically, record the filename in the notebook.
- Record observations made during data collection. Anomalous observations may indicate limitations of or errors in the hypothesis.
- Record errors, problems encountered, how they were corrected, and their possible impact on the results.
- Record all sample calculations.
- Record any thoughts or perceptions that might be relevant to understanding the experiment.
- Record discussions with colleagues and supervisors.
- Once you have finished an experiment or project, summarize that experiment in your notebook. These summaries form the basis for reports, publications, and presentations.

continued on page: end

Signature	Date	Witness	Date
Roy Jensen	20 June		

5.2 Storyboarding

It is all-too-common for people to sit down in front of a computer or with a pad of paper and have their mind go blank. They know the goal is to answer the research/investigative question and they have a good understanding of the information and what needs to be conveyed, but the big questions are "Where do I start?", "What do I say?", and "How do I say it?" Much of this hesitation comes from having too much information, from not knowing what information must be conveyed, from having too long a document to prepare, and from believing they are writing the finished document. For you to avoid and overcome these barriers,

- *start early*: by distributing the writing process throughout the project timeline, you are not overwhelmed by the volume of information you have or the volume of writing you must do.
- *realize you are not writing the finished document*: once you accept that what you write does not need to be perfect — that it will require review and revision — it is easier to write material that will eventually transform into the finished document.
- *prewrite*: prewriting is the process of collecting and organizing information to establish the structure and content of your document.

Prewriting

There are many methods of collecting and organizing information. Below are some common prewriting strategies. You may find that you use all of these strategies at different times, but that one generates the most information and smoothly leads to the draft document.

- *Brainstorming*: writing the ideas and information that is relevant to the project in point-form. Supporting information is not required, just enough to remind you so that you can you can fill in the details and place the material in context later. The goal is to get as much information onto paper as possible. Often, writing an idea triggers you to think of more. The information is subsequently organized into sections and detail is added to form the draft document.
- *Freewriting*: similar to brainstorming, but information is written as sentences and the goal is to provide detail on each topic. Freewriting encourages you to jump to another topic if your mind switches to it. Punctuation and grammar are not important, nor are the links between topics. Once complete for all of the topics, the sections are organized to form the draft document.

- *Questioning*: consider the project as a news story and ask questions about the project. Strive to answer the questions *who*, *what*, *where*, *when*, *why*, and *how* about the project. These questions may identify better questions. The answers form the draft document.
- *Storyboarding* (also *outlining*, *mapping*): a more structured version of brainstorming where the information is organized directly into sections. Each section is developed independently of the others, and the combined sections forms the draft document.

Storyboarding

Storyboarding is commonly used when preparing scientific documents and presentations because of the established common structure in scientific documents. For example, the *Introduction* can be storyboarded as part of the literature survey. Storyboarding then continues during the project and evolves into the complete document. This is consistent with Figure 4.2 (page 127), which shows that writing is part of the research process.

There is no explicit method for how a person should storyboard a document. Some people storyboard on paper with separate pages for each section, as illustrated by Figure 5.3. Others develop the document dynamically in a word processor. You need to try different methods and find a method that works for you.

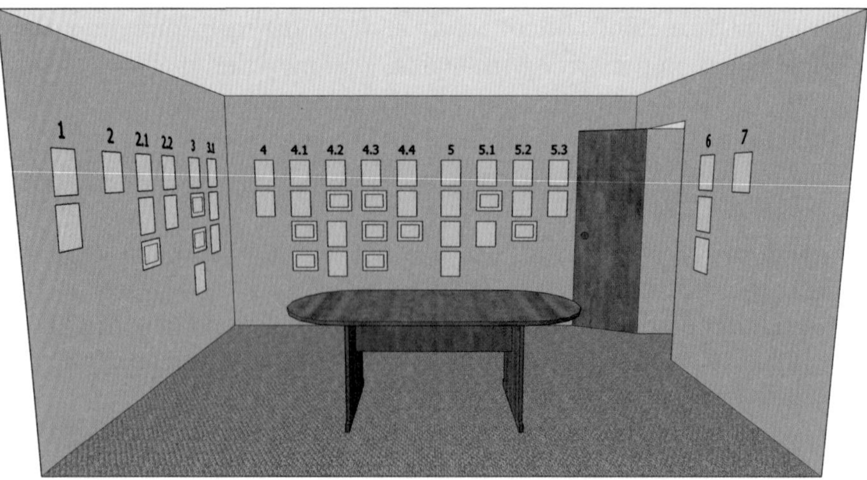

Figure 5.3 Storyboarding a document by posting sections on walls and organizing the information into those sections. Note that the most technological of organizations — movie studios — still storyboard on paper!

Regular review and reorganization of the information allows you to draw connections between data, link your results to what is already published, and identify areas where more research/investigation is required. The flowcharts in sections 5.6 to 5.9 contain a storyboarding section similar to Figure 5.4.

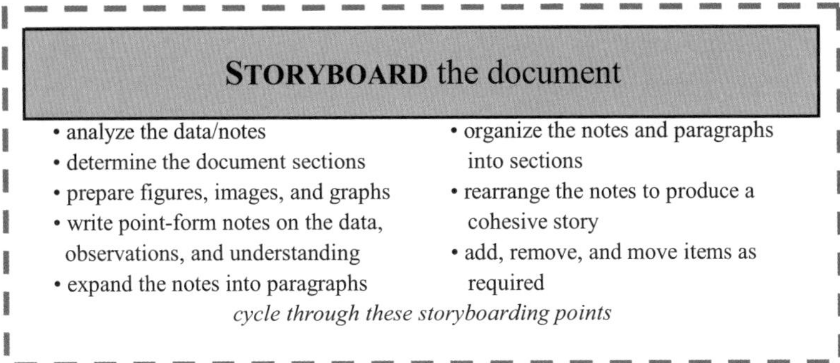

Figure 5.4 The process of storyboarding.

5.3 Laboratory reports

The laboratory report is the first document produced after an experiment is conducted. Depending on what you are doing, you will usually prepare a *complete report* or a *summary report.*

A ***complete report*** is prepared the first time an experiment is conducted. It contains all the sections detailed below. Complete reports are required for validation, accreditation, and to teach new persons how and why an experiment is done in a particular manner. The complete report forms the basis for the report given to a client and is the starting point for scholarly articles. For example, forensics laboratories prepare complete reports to validate their results; lawyers target the methods used in forensic laboratories with the hopes of discrediting the results.

A ***summary report*** is prepared when the analysis is routine. That is, a complete report has already been done for the experiment and only different samples are being tested. The summary report refers to the complete report for the theory, materials, and procedure and focuses on the results, discussion, and conclusions from that particular experiment. Examples include medical tests, weather reports, and environmental reports.

In addition to laboratory reports, science writers commonly prepare *progress reports* and *project reports*. ***Progress reports*** are interim reports prepared as part of large research and investigative projects. ***Project reports*** are reports prepared at the end of major projects that involve many experiments.

Sections in a laboratory report

The following sections are commonly used to organize laboratory reports. The section headings you use will depend on your discipline, the nature of the experiment, and the preferences of you and your organization. It may be possible to modify these headings to make them more informative and/or add sections if appropriate for the discipline or experiment.

- *Abstract*
- *Introduction*
- *Theory*
- *Materials*
- *Methods* or *Procedure*
- *Hazards*

- *Results*
- *Discussion*
- *Conclusion* or *Summary*
- *Bibliography* or *References* or *Works cited*

Some instructors, employers, and publishers prefer the sections numbered. Sections may also be combined: *Introduction and theory*, *Materials and methods*, *Results and discussion*. The *Hazards* section can be integrated into the *Materials* section when the hazards pertain to the materials themselves, such as chemicals or equipment, and the combined materials and hazards may be tabulated. Hazards associated with the procedure can be integrated into the *Methods* section.

The *Abstract* is a concise overview and summary of the theory, experimental method, and key results. The *Abstract* must itself be a complete document as it is often read separately from the entire report.

A well-written *Introduction* informs the reader that you understand the nature of the experiment and how it fits into the bigger scientific picture. A detailed *Theory* section indicates you understand the science behind the experiment. Detailed *Results* and *Discussion* sections illustrate your ability to analyze and interpret the data in the context of the underlying theory, and convey its importance to science and society.

When preparing a laboratory report, it is often valuable to start with the section you are most comfortable with — often the *Materials and methods* or *Results* section — and then move to other sections. You will find that as you write, your understanding and ability to explain other sections improves. Expect to move between sections as you complete the report.

Progress and project reports

In addition to laboratory reports, common reports include *progress reports* (status reports, weekly updates, group meetings) and *project reports* (feasibility studies, technical reports, comparative assessments, environmental impact reports, long-term planning projects, final reports).

Progress reports

Large research and investigative projects often require the investigators to submit regular progress reports. These reports provide the supervisor with the status of individual projects to compare with the status projected in the research proposal. The supervisor can identify potential problems and adjust resources (staff, funding) accordingly. Progress reports by the supervisor provide those funding the research with the status of the overall project.

Progress reports also provide the writer — from the student working on a term essay, to the undergraduate or graduate student involved in research, to the principal investigator of the research program — with an opportunity to review and reflect on the work completed. Future work can be planned to complete the project more efficiently. Progress reports also serve as a starting point for preparing other documents.

Progress reports to a supervisor are typically short, one to three pages, and could alternatively be in the form of an oral presentation. Progress reports to funding organizations are typically longer.

Progress reports often form the basis for the project report, which draws the results of the individual experiments into a coherent document for the reader.

Project reports

Large research projects are common in universities and in industry, while large investigative projects are common in university, industry, and government. For example, science graduate students design and conduct a research project under the direction of a faculty member. Each student's project is part of the faculty member's larger research program.

Project reports will be read by a broad audience: scientists, engineers, managers, accountants, government officials, reporters, and the public. Quite often, these people will not read the entire report, focusing instead on the section(s) that is interesting or important to them. Thus, every section of the report should be reasonably self-contained so that it can be understood without reading the entire report.

Formatting reports

Laboratory and progress reports are produced with minimal formatting as illustrated below. See page 146 for guidelines on how to format these documents.

Project reports are formatted similarly to a thesis (for a research project) or essay (for an investigative project). See Section 5.6 for details on preparing these documents.

Reviewing reports

You want your report to be free of errors, easy to understand, and interesting. Appendix A lists questions you should keep in mind when preparing your report. Additionally. ensure the readability statistics are appropriate for the audience (see Table 5.1). Once you have prepared your report and reviewed it once or twice,* have one or more colleagues review it and provide feedback. Give reviewers the Appendix A questions and ask them for recommendations to improve the report. Chapter 6 provides more information on the review process.

* Place the report aside for a day, or at least a few hours, and then reread it. You may be surprised that what made sense when you wrote it does not make sense now. Revise the report appropriately. (See page 131 and section 6.1 for details.)

... laboratory report ...

A clear, concise title that accurately reflects the content of the report and is understood by a broad scientific audience

name, date

Abstract

A concise overview and summary of the theory, experimental method, and key results. The *Abstract* should place the current experiment in the context of the greater research project, if applicable. The *Abstract* must itself be a complete document as it is often read separately from the entire report.

(up to $^{1}/_{4}$ page)

Introduction

An overview of the experiment conducted, the objective(s) of the experiment (often in the form of a research question), the rationale for the experiment being conducted, and relevant background information to put the experiment in the context of the larger project. The introduction sometimes ends with a one-sentence summary of the conclusion, "I found that ...".

($^{1}/_{2}$ – 2 pages)

Theory

The theoretical foundation underlying this experiment. From the theory, one can predict the experimental outcome.

($^{1}/_{2}$ – 1 page)

Materials

The equipment and supplies used to conduct the experiment. For equipment, include the make and model and any modifications. For supplies, include the manufacturer and lot number.

(as much space as required)

Hazards

Document all known hazards regarding the materials and procedures for the safety of others who may repeat your work. Indicate the appropriate handling and safe-use procedures. If something potentially hazardous happened during your experiment (explosion, fire, accidental release, mechanical failure, ...), document this and include strategies for minimizing the risk of similar accidents in the future.

(as much space as required)

– 1 –

... laboratory report ...

Method *or* **Procedure**

A detailed description of how the experiment was conducted, including instrument settings during data collection. A schematic of the experimental setup aids in understanding. The procedure must be complete enough that anyone with similar training can conduct the experiment.

Note: a laboratory manual lists the steps necessary to complete the experiment; a *Methods* section is written in paragraph form as a process.

($^1/_2$ – 2 pages)

Results

This section presents the data obtained and the analysis of that data. Data are commonly presented in tables and graphs; use chemical and/or mathematical formulae to explain how the results are derived from the data.

($^1/_2$ – 2 pages)

Discussion

This section presents a detailed interpretation of the results in the context of the research question. This is the most important section of the report, as understanding is only achieved when meaningful information is extracted from the data.

($^1/_2$ – 2 pages)

Conclusions

This section summarizes the key points from the results and discussion and relates the results to the experiment objectives.

(up to $^1/_2$ page)

Bibliography *or* **References** *or* **Works cited**

A list of the resources you used to prepare the report. Common references for laboratory reports include scholarly books and scholarly articles. In schools, the laboratory manual and textbooks are often cited.

(as much space as required)

Appendices

This section contains supplementary information not directly pertinent to the objectives of the report but of value to some readers. Appendices often contain the raw data (if not in the *Results* section), the derivation of formulae, sample calculations, and secondary analyses. Ensure the report contains links to each appendix.

– 2 –

... progress report ...

A clear, concise title that accurately reflects the project and is understood by a broad scientific audience

Progress report to <date>

name

Introduction

An overview of the project that is a summarized version of the *Introduction* in the original proposal (section 5.5). The *Introduction* should focus on the component(s) (experiments or investigative steps) of the project that you are working on. The introduction does not change between progress reports unless there is a change in the project plan. ($^{1}/_{4}$ – 1 page)

Results

For each component that you are working on, focus on the work since your last progress report.

- Summarize the work completed to date. Compare the work completed against the schedule and budget in the proposal.

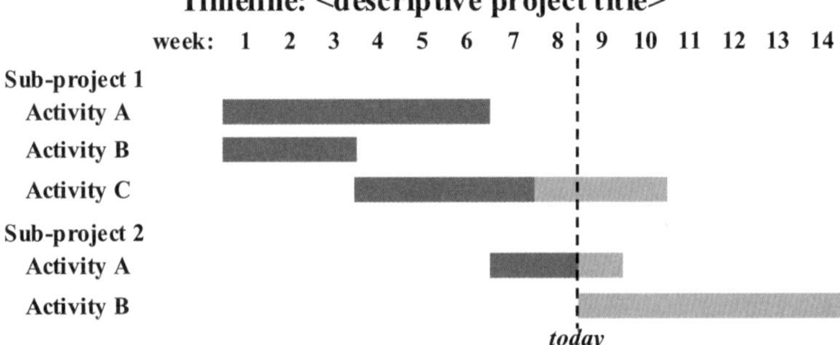

- Discuss successes.
- Discuss problems encountered and how they were/are being addressed.
- List the work remaining in each component and what you plan to do before the next progress report is due.
- Include recommendations if changes are required.

As a component nears completion, summarize the results obtained (tables and graphs) and how those results affect the remainder of the project. The final progress reports should become increasingly like a laboratory report for your component of the larger project. Introduce sections such as a *Discussion*, *Conclusion*, and *Future work*. (1 – 3 pages)

– 1 –

5.4 Article summaries

As you progress through your education and career, you will need to read and summarize the work done by others. An ***article summary*** (***précis***) is a $^{1}/_{2}$- to 2-page document that summarizes the important aspects of a scholarly article, usually following the same order and logic as the article. The summaries provide you with a shorter document to review when refreshing your memory of the article. Students may also be required to submit an ***annotated bibliography*** as part of a term essay. An annotated bibliography is a collection of article summaries.

An article summary is not an extraction of key sentences into another document. A summary requires you to understand what the scientists have done and what knowledge they obtained from their research, and explain it in your own words. A summary extends beyond the reported work to include your opinion of the work and its importance to your project. Your summary should answer several questions:

1. Why was research conducted?
2. What research was conducted? (the research question)
3. What are the benefits and limitations of the research method?
4. What knowledge was gained from this research?
5. What questions are still unanswered?

Question 5 could easily set the stage for your own research project!

Sections in an article summary

The following sections are common in summaries, but variation is also common depending on the nature of the article and the information you are interested in.

- *Overview*
- *Research method*
- *Research results*
- *Personal observations*

These sections may be written in paragraph and/or point-form.

Your summary must be comprehensive of the entire article. When you are primarily interested in one aspect of an article, the summary should still contain a brief overview of the article but focus on the aspect you are most interested in.

Flowchart for preparing an article summary

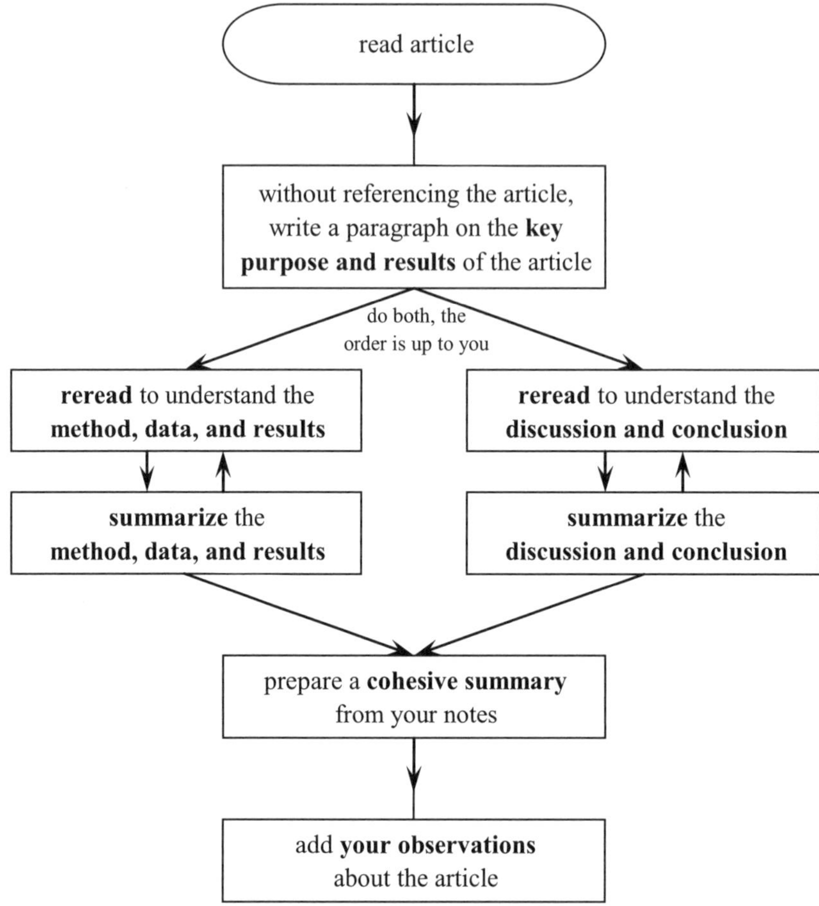

Figure 5.5 A flowchart outlining the steps to preparing a research article summary.

In summarizing the article, ensure that your summary is accurate and that you are not simply extracting sections of text from the article. While Figure 5.5 shows that summarizing occurs with concurrent reading of the article, you may find it easier to write the summary in your own words without referring to the article. Avoid quotations unless there is a profound or exceptionally concise statement that you cannot further summarize.

You may find that your first draft is longer than two pages. As you prepare subsequent drafts, focus on the important information from your earlier drafts to increase the clarity, cohesion, concision, and precision of your summary.

Formatting an article summary

Article summaries are produced with minimal formatting. See page 146 for guidelines on formatting these documents.

Depending on your needs, it may be better to prepare the summary in a citation management program that stores the complete reference, a copy of the article, and your summary. Additionally, all the text is fully searchable. See page 138 for details.

Reviewing article summaries

Article summaries are not normally published. However, they may be assigned as course assignments or be required in an annotated bibliography.

You want your summary to be free of errors, easy to understand, and interesting. Once you have completed the summary, place it aside for a few hours or days, and then reread the article and your summary. Revise as required to ensure your summary is complete.

If you are submitting the summary, have a colleague review your summary and provide feedback. Give the reviewer the article and the Appendix A questions and ask them for recommendations to improve the summary. Chapter 6 provides more information on the review process. Additionally. ensure the readability statistics are appropriate for the audience (see Table 5.1). Revise the summary appropriately.

...article summary...

A review of
<complete citation of paper>

name, date

Overview

The first sentence should present the key result(s) of the article.

The remainder of the *Overview* should summarize the theory, explain why the research was conducted, and place the research in the context of the greater scientific field.

(up to ½ page)

Research method, data, and results

A summary of the research conducted and the results thereof. While these may be in separate sections in the original article, they should be blended together in the summary for concision.

The last paragraph in this section could be your observations on the method and analysis, or this could be included in the *Personal observations* section.

(½ – 1 page)

Discussion and conclusions

A summary of the knowledge gained from the research and the importance, benefits, and limitations of the research.

The last paragraph in this section could be your observations on the discussion and conclusions, or this could be included in the *Personal observations* section.

(½ – 1 page)

Personal observations

Your professional opinion of the article: what was good, poor, and interesting about the article? Is there any potential bias based on the author, funding source, or publisher? What other limitations did the article authors not discuss? What unanswered questions are there; what could be the next step in this research; and how could these be answered or conducted? Write these comments in a positive tone.

(up to ½ page)

– 1 –

5.5 Research proposals

Money is required to conduct and disseminate research, from purchasing equipment and supplies, to paying researchers and assistants, to travelling and presenting at conferences. Scientists apply for funding from government, private, and corporate organizations. In every application, the scientist must make a persuasive argument for why they deserve funding.*

A ***research proposal*** is a sales pitch to funding agencies for money to conduct research. The research proposal must show that you have conducted a thorough literature survey and have thoroughly thought out the research you propose to do. In essence, you must sell your research as a valuable advancement of science and show that it will benefit society. A panel of scientists in the same field reviews the proposals — nuclear physics proposals are reviewed by nuclear physicists, etc. — so your proposal must be targeted to specialists in the field.

Most funding organizations receive more applications than they have funds for, so the reviewers are looking for reasons not to fund your research! Your application must be clear, coherent, concise, precise, and *persuasive*, and the project must be within the funding mandate of the organization. You need to sell your idea, your abilities, your confidence, and your enthusiasm to the reviewers.

How you prepare your proposal depends on the audience. Proposals prepared for government funding organizations should focus on how the work will advance science and improve understanding. Extravagant claims and grandstanding will be questioned. Proposals prepared for private and corporate organizations should promote the accomplishments of the researcher(s). Be aware that private and corporate organizations (and sometimes governments) may have agendas and you must tailor your proposal accordingly. Your application may propose methods for obtaining the preferred results, but you must always avoid engaging in scientific misconduct. See Section 4.2 for more information on research ethics.

* Think of a research proposal as a specialized form of a business plan. Whereas a business plan is scrutinized by bank loan officers and investors who want to ensure they will get their money back with interest, a research proposal is scrutinized by other scientists to ensure the money will result in an advancement of science.

Your proposal is a legally binding document. You are not agreeing to get the results you propose, but you are agreeing that you have the ability to complete the project and will commit the resources, personnel, and funding to the project.

When applying for funding, you must use the application form of the organization to whom you are applying; your application must contain all the required information; and you must follow their guidelines regarding format and length. Recall that the reviewers are looking for reasons *not* to fund your project. Sample forms can be found at

NSERC:	http://www.nserc-crsng.gc.ca	(Canada)
SSHRC:	http://www.sshrc-crsh.gc.ca	(Canada)
NSF:	http://www.nsf.gov	(United States)
NIH:	http://www.nih.gov	(United States)

The template research proposal below contains sections that are commonly required by funding organizations.

Reviewing research proposals

You want your proposal to be free of errors, easy to understand, and interesting. Appendix A lists questions you should keep in mind when preparing your proposal. Additionally, ensure the readability statistics are appropriate for the audience (see Table 5.1).

Once you have prepared your proposal and reviewed it once or twice,[*] have one or more colleagues review it and provide feedback. Give reviewers the Appendix A questions and ask them for recommendations to improve the proposal. Chapter 6 provides more information on the review process.

Student scholarships

Scholarships are a form of proposal in that you must *persuade* the scholarship administrators that you are the best candidate to receive the scholarship.

[*] Place the proposal aside for a day or two, and then reread it. You may be surprised that what made sense when you wrote it does not make sense now. Revise the proposal appropriately. (See page 131 and section 6.1 for details.)

... research proposal ...

A clear, concise title that describes the proposed research, the investigated variables, and the research methodology

name, date

Collaborators

The scientists who will be working with you on this research project. List their qualifications and the resources they have available to contribute to the project.

Summary

A plain-language summary targeted to the public, focusing on the societal impact of the proposed research. Prepare the *Summary* as a freestanding document, because it will be published separately if the research is funded.

(up to $^1/_2$ page)

Abstract

A technical summary targeted to the scientific reviewers. It should include the research question, the hypothesis, and summary of the research design and expected outcomes. Reviewers often read the *Abstract* first, so it should be clear and persuasive.

(up to $^1/_2$ page)

Introduction

A review of the relevant scientific literature in this area, limitations of the previous research, and how your proposed research builds on the previous research. Show how your research is *significant* and *important* — remember you are trying to sell yourself! Indicate how this research fits into your research interests and other research projects. This section should cite numerous peer-reviewed sources, including yours if you are active in this area. Avoid vague statements and statements not supported by the literature.

The literature review is the most critical part of a research proposal. It is also often the most lacking. A thorough literature review shows that you have spent time learning about this area of science, demonstrates your knowledge of the historical and current work in this area, demonstrates your ability to critically evaluate the literature, and convinces the reader that you are serious about conducting the research you propose.

(1 – 2 pages)

– 1 –

... research proposal ...

Research plan

A detailed description of the research you wish to accomplish, the research question, research design (the types of experiments to be conducted, the source of samples, the equipment to be used, and how the data will be analyzed), expected practical and/or theoretical outcomes, and the strengths and limitations of your research plan.

(1 – 2 pages)

Facilities and equipment

List the facilities and equipment required to conduct the experiment, and their location. If the facilities or research site are not at your institution, attach documents confirming permission to use the facilities or access the site. (This is not required if the facilities are affiliated with a collaborator. If facilities must be reserved (such as ship time or satellite time), attach documents confirming pre-approval to reserve the facilities.)

(up to 1 page)

Ethical approval

If the research involves living organisms, including human subjects, the proposal must include proof that an independent Institutional Review Board has reviewed and approved the research plan (see section 4.2).

(attached as appendix)

Schedule

A timeline for the research: obtaining equipment, set-up and testing, data collection, data analysis, and reporting. If there are multiple scientists involved and the project has multiple facets, identify who will be doing which facet of the research. Use a chart to visualize the timeline.

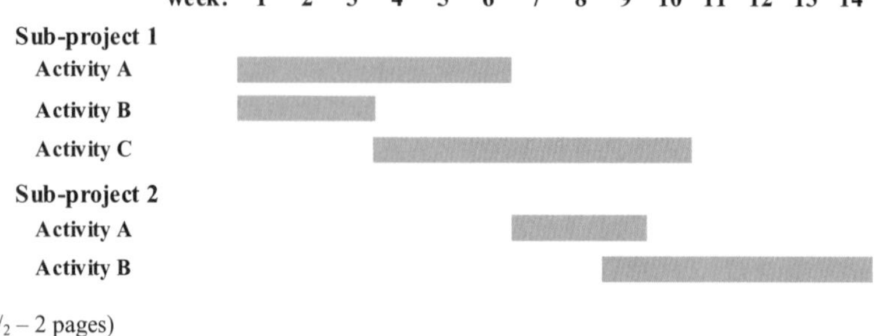

(1/2 – 2 pages)

– 2 –

... research proposal ...

Budget

A detailed list of the resources required to complete the research: salaries, equipment, materials, travel costs for fieldwork and meetings, and publication costs (conferences, etc.). List the funding expected or obtained from other sources and the funding requested from this organization.

($^1/_2$ – 1 page)

Benefits

Detail the expected benefits and/or deliverables of the research to the funding agency, to the scientific community, and to society. Also, detail the expected number of researchers and student researchers who will be involved in the research project.

($^1/_2$ – 1 page)

Summary

A closing paragraph that summarizes the importance of your proposed research and emphasizes your enthusiasm and confidence in successfully completing the work.

($^1/_2$ page)

Bibliography *or* **References** *or* **Works cited**

A list of the resources you used to prepare the proposal. Most of these should be scholarly articles.

(as much space as required)

– 3 –

5.6 Scholarly articles, including essays and theses

A scholarly article[*] is the primary means of communicating your research to other scientists. Scholarly articles expand human knowledge as illustrated in Figure 1.4 (page 9). By their nature, they are highly technical documents.

A fundamental requirement of scholarly articles is that the articles undergo a rigorous peer review process prior to publication. Scholarly articles are reviewed by scientists who are independent of your research but who are knowledgeable in the field. The reviewers have significant power: they may recommend publishing the article as is, publishing with minor modifications, recommend additional experiments be conducted, or may recommend not publishing the article. The goal of this rigorous peer review is to ensure that the published work is of the highest quality.

Types of scholarly articles

Research articles are the most common type of scholarly article. They are prepared after the completion of a research project and provide a comprehensive report of that research to the scientific community.

Communications are shorter scholarly articles that quickly communicate profound results to the scientific community.

Review articles review and summarize what has been published in a scientific field and endeavor to summarize the current understanding of that field. It is common for review articles to be 50 or more pages and cite hundreds of references.

An *essay* (or *report*[†]) is the culminating document from an investigative project. Essays are routinely prepared for non-specialists in the field, such as executives, politicians, funding organizations, and the public. In an academic setting, essays are prepared as a course requirement.

A *thesis* or *dissertation* is the culminating document that presents the research conducted by a student. Students who take an undergraduate research course as part of their degree (B.Sc., B.Eng.) may write a thesis. Students in a masters (M.Sc.) or doctoral (Ph.D.) program must write and defend a thesis before receiving their degree. The difference between an undergraduate, masters, and doctoral thesis is the amount of independence the student has and the volume of original research required of the student. A doctoral thesis must make a substantial

[*] Also called *papers*, *primary literature*, and, when in development, *manuscripts*.

[†] Not to be confused with a laboratory report.

contribution of new knowledge, the others decreasingly so. What defines substantial? Your supervisor and defense committee decide that.

Storyboarding a research article or communication

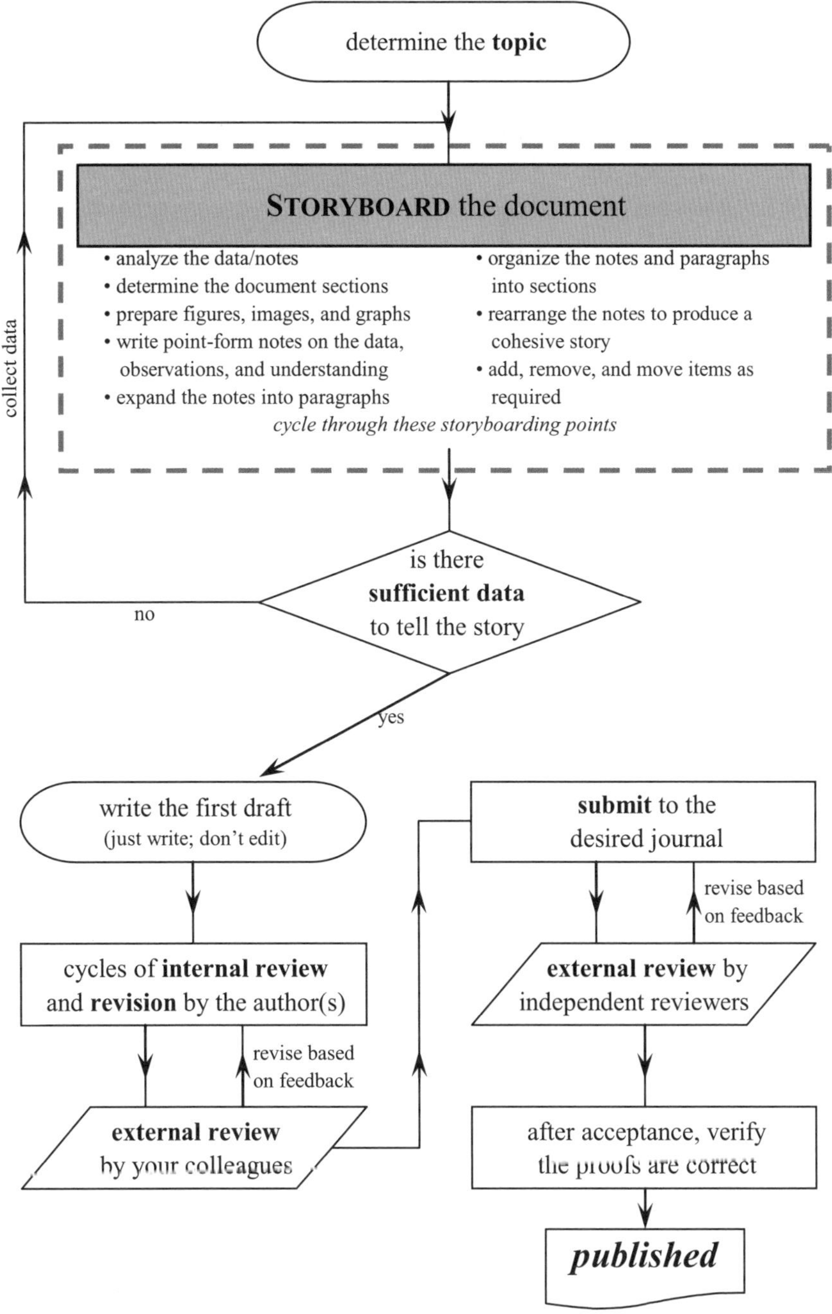

Figure 5.6 A flowchart outlining the steps to preparing a research article or communication.

The first draft is exactly that: a draft. The goal of the first draft is to collect all the information in one location and make obvious connections between the points. It does not need to be neat. Subsequent reviews by yourself and your colleagues correct the grammar, style, and add depth to the work.

Sections in a research article

Every scholarly journal has exacting specifications that must be followed when submitting an article to the journal. The recommendations below are representative of the journal requirements for research articles.

Like laboratory reports and research proposals, research articles have standard sections. It may be possible to modify your section headings to make them more informative and more appropriate to the discipline, article, and/or journal.

- *Title*
- *Author* and *author affiliation*
- *Abstract*
- *Introduction*
- *Materials and methods* or *Procedure*
- *Results*
- *Discussion*
- *Conclusion* or *Summary*
- *Appendices*
- *Bibliography* or *References* or *Works cited*

The *author list* is the names of people who have made intellectual contributions to the research. Write names as first name, initials, and surname: Roy H. Jensen. Do not include professional or official titles or academic degrees. Authors are usually listed in order of decreasing contribution to the research, with the principal investigator listed last. One author, usually the principle investigator, is designated as the author to whom correspondence should be addressed. For this author, the full mailing address, phone number, fax number, and email address is given, usually in a footnote. The *affiliation* of each author is the institution and department of the author. If multiple affiliations, label with superscript a, b, c, … or *, †, ‡, ….

The *Title* and *Abstract* have a very important role. Together, they must summarize the research question, research design, and key results. The *Abstract* should explain why the research is interesting, important, and relevant, and how the research fits with and expands our understanding of this scientific field. The *Abstract* also explains what research was conducted and the key results of that research. The *Abstract* must be a complete document, as it is published separately. On the internet, often only the title and abstract are available to readers without a subscription.* They are also the only sections returned in a database search. The *Title* and *Abstract* must summarize and promote the document, convincing the reader that the full document is worth reading.

Because abstracts must be short, authors often find writing the *Abstract* difficult. Many leave it to the end, after the article has been written and the authors have a thorough understanding of what is included therein.

Omit extraneous text and superfluous phrases in the *Title* and *Abstract*. For example, remove phrases like "A study of ...", "Investigations of ...", "Observations on ...", etc.

Sections in communications and review articles

Because communications are shorter documents meant to convey profound results, they typically focus on the *Results* and *Discussion* sections and only have a brief *Introduction* and *Procedure*, referring readers to their future research article.

The traditional headings are not applicable to review articles because the intent is not to present research, but to summarize the research conducted by many scientists. Persons writing review articles have more flexibility in determining section headings and often use headings that are descriptive of the information presented in the section.

Formatting an article

Different scientific disciplines have different preferred formats for scholarly articles. Even within a given discipline, different scientific journals have different style guides. Authors must follow the style requirements of the journal they are submitting to! Some journals require a document that is double-spaced with minimal formatting. Other journals require the author submit the scholarly article in publication format. Thankfully, journals that require submission in publication format provide an electronic template to facilitate this. However, configuring your word processor to simulate publication format teaches you lot about using your word processor.

* Pay for articles only as a last resort! You can often get articles free through inter-library loan, from the corresponding author, or from a local university.

Page 146 has general guidelines to format documents. Specific formatting unique to scholarly articles is given below:

- *text*
  - 1.0 spacing (single spaced)
  - two columns with a 1.0 cm spacing
- authors and their affiliations as shown on page 177
- submission timeline as shown on page 177

A scholarly article formatted for publication is presented below.

Reviewing scholarly articles

A key feature of publishing scholarly articles is the peer-review process. If you publish scholarly articles, you are required to review scholarly articles published by others. Expect to review two to four scholarly articles for every article you publish. This obligation is part of your commitment to ensuring that only quality articles are published. This obligation extends to reviewing the work of your peers as you proceed through undergraduate and graduate school.

Appendix A lists questions you should keep in mind when preparing your article. Additionally, ensure the readability statistics are appropriate for the audience (see Table 5.1). Once you have prepared your article and reviewed it once or twice,* have one or more colleagues review it and provide feedback. Give reviewers the Appendix A questions and ask them for recommendations to improve the article. Chapter 6 provides more information on the review process.

* Place the article aside for a day or two, and then reread it. You may be surprised that what made sense when you wrote it does not make sense now. Revise the article appropriately. (See page 131 and section 6.1 for details.)

… formatted for publication …

A clear, concise title that accurately reflects the content of the article and is understood by a broad scientific audience

Jessie A. Wang,[a] Jamie B. Martinez,[a] Taylor C. Müller,[b] Riley D. Jensen[a,†]

a. *institution and department of these authors*

b. *institution and department of this author*

Received <date>; revised <date>; accepted <date>

A concise overview and summary of the research question, research design, and key results. The *Abstract* should inform the reader why the research is interesting, important, and relevant; how the research fits with and expands our knowledge in this field; what research was conducted; and the key results of that research. The *Abstract* must itself be a complete document as it is also published separately.

Introduction

A review of the relevant scientific literature in this area and how the current research — the subject of this scholarly article — builds on the existing research. This section should cite numerous peer-reviewed sources.

The *Introduction* should answer four main questions: What is already known in this area? What research was conducted? Why was the current research conducted? What are the key results of the current research? Answering this last question foreshadows the results.

The goal of the *Introduction* is to convince the reader that the current research is an important contribution to the scientific field.

†. Corresponding author: <mailing address, email address, phone number>

Materials and methods

Write this section as a process: each step in order. It must be sufficiently complete so that anyone with similar training is able to conduct the experiment. A schematic of the experimental setup aids in understanding. If the procedure has been published, reference it, but still give a brief description and detail any changes.

When documenting a procedure for the first time, document each step, including the equipment and supplies used to conduct the experiment. For equipment, include the make and model of the equipment, any modifications to the equipment, and the instrument settings during data collection. For supplies, include the manufacturer, lot number, and how the supplies

... formatted for publication ...

were stored or handled prior to use in this experiment. Document all known hazards regarding the materials and procedures. Indicate the appropriate handling, safe use, and safe disposal procedures.

Results

Present the data obtained and the analysis of that data. Data is commonly presented in tables and figures, but do not present the same data in both a table and a figure. Chemical and/or mathematical formulae explain how the results are derived from the data.

While laboratory reports and theses require all data be presented, research articles and communications only contain representative data and calculations. (This saves publication space.) The journal may require all the data be submitted electronically so that it is available to others.

Readers accept that the analysis of the remainder of the data is similar to the sample shown. Explicitly state if some data or results are inconsistent with the majority of the data.

Discussion

This section presents a detailed interpretation of the results in the context of the research questions and relates this new information to existing knowledge in the field. The discussion should be thorough, but concise. Avoid repetition of information presented elsewhere. Ensure that the statements are supported by the results!

Be sure to discuss the theoretical and practical applications of this new knowledge, and the significance to science and society.

Conclusion

Summarize the key points from the *Results* and *Discussion*.

Acknowledgements

Acknowledge those who provided assistance but are not authors on the document: those who provided samples, technical assistance, advice, ... Acknowledge organizations that funded the research.

Disclose any potential conflicts of interest.

Works cited

A list of the resources you used to prepare the article. Most of the resources should be peer-reviewed scholarly articles and scholarly books.

It is critical to double-check the citations; errors are surprisingly common.

Appendices

Appendices contain specialized information not critical to the article, but important to those conducting similar research. This includes derivations, unique mathematical calculations, and rationalizations for the procedure or analyses conducted.

– 2 –

Storyboarding an essay or thesis

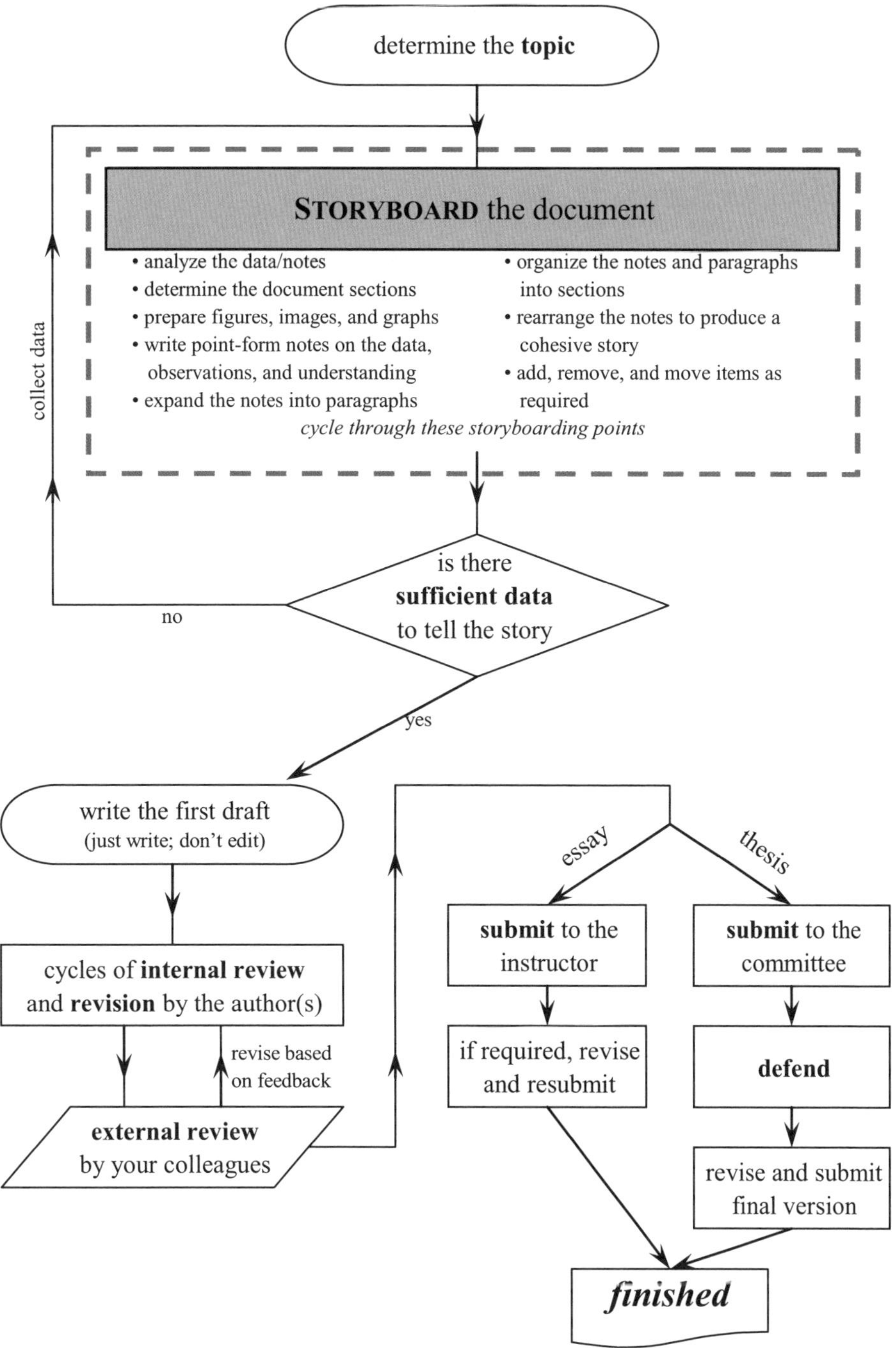

Figure 5.7 A flowchart outlining the steps in preparing an essay or thesis.

The first draft is exactly that: a draft. The goal of the first draft is to collect all the information in one location and make obvious connections between the points. It does not need to be neat. Subsequent reviews by yourself and your colleagues correct the grammar, style, and add depth to the work.

Sections in an essay or thesis

Essays and theses are formatted like a book. In a thesis, the section headings are similar to a research article, but draw together several experiments as part of the overall research project. Institutions provide guidelines and templates for students preparing theses. In an essay, the experimental chapters are replaced by chapters exploring the research conducted by others.

The goal of an essay/thesis is to logically and coherently draw together the existing knowledge and current research — including your own research, if you are writing a thesis — to increase the knowledge in the field. Essays and theses do not simply summarize the individual works, they analyze them to find commonalities and discrepancies to better understand the field. The discrepancies indicate areas where understanding is poor and could form the basis for future research projects.

There is often a word limit on essays. This limit refers only to words in the numbered chapters, *Introduction* to *Acknowledgements* (see below). Theses do not have a defined length — they thoroughly but concisely present all the research that was conducted. On average, the *Introduction* to *Acknowledgements* chapters of undergraduate theses are 20 to 50 pages, Master's theses are 80 to 200 pages, and Doctoral theses are 150+ pages. *Appendices* could add hundreds of pages, depending on the nature of the research.

Formatting an essay or thesis

Essays and theses are produced with minimal formatting. Page 146 has general guidelines to format documents. Specific formatting unique to essays and theses is given below. However, if you are preparing a thesis and your department or institution has guidelines, those guidelines supersede these guidelines.

- title page with *Title*: 20 point font; bold
- *headings*
    - Headings may be numbered in longer documents.
    - Essays and theses may require additional heading levels:

        Heading 1: 16 point; bold; numbered; prefaced with "Chapter"; space after: 18; start on new page

        Heading 2: 14 point; bold; numbered; space before: 18; space after: 6

        Heading 3: 12 point; bold; space before: 12; space after: 3

Alternatively, essays and theses may be formatted like a book.

Below is an example of an essay and of a thesis formatted for submission. Each section details the information expected in that section. The essay is assumed to be a student essay submitted as part of a degree program. However, as noted on page 158, essays are commonly prepared for other purposes such as project reports. In these circumstances, sections unique to student essays are not included.

... essay template ...

A clear, concise title that accurately reflects the content of the essay and is understood by a broad scientific audience

by

<author>

in partial fulfillment for the requirements of <course>

at

<institution>

<date>

... essay template ...

Abstract

A concise overview and summary of the problem you are investigating and the key results. The abstract should inform readers why the investigation is interesting, important, and relevant; how the investigation fits with our knowledge in this field; and the key results of the investigation.

The *Abstract* must itself be a complete document as it is also published separately.

(The *Abstract* is on its own page. 100 – 300 words)

Table of contents

A table of contents that lists the chapters, headings, and numbered sub-headings and their page numbers. Appendices are included in the *Table of contents*.

Appendix B of *Communicating Science* explains how to use styles, which will dynamically create and update your *Table of contents*.

See, for example, the table of contents of *Communicating Science*.

(The *Table of contents* is on its own page(s).)

The above sections either can be without page numbers or numbered using roman numerals. If using roman numerals, the title page is page i, the *Abstract* is on page iii, and the *Table of contents* on page v (assuming a double-sided document). However, do not show the page number on the title page.

Executive summary

A longer overview of the report with the technical details minimized and the results and recommendations emphasized. The format and headings of the *Executive summary* are similar to the overall report, including the repetition of important tables and figures. Emphasis is on the importance, benefits, conclusions, and recommendations of the project. The *Executive summary* is typically 5 – 10 % of the total report length.

(The *Executive summary* is on its own pages and numbered using roman numerals.)

1. Introduction

This section introduces the topic of the essay, why you have chosen this topic, and what you intend to show or argue (foreshadowing your conclusions). Often, the *Introduction* identifies a problem and then explains how the essay will address the problem.

The goal of the *Introduction* is to convince the reader that the essay makes important contributions to understanding the topic and field.

(The *Introduction* starts on page 1. 10 – 20 % of essay.)

– 2 –

... essay template ...

2. <Information from literature: source 1>

This chapter presents the information you discovered during your investigation. You must present the information in a clear and logical manner so that the reader can follow your logical progression to the conclusions you plan to draw later in the essay.

Since the information comes from other sources, this section must cite these sources.

(Each information chapter is 10 – 20 % of essay.)

3. ... <Information from additional literature sources>

Repeat the steps in Chapter 2 for each source.

7. Discussion (assuming chapters 2 – 6 explore different literature sources)

This section draws connections between the information presented in previous chapters and explains how it addresses the problem identified in the *Introduction*. You can also explore the broader implications of your investigation to science and society.

The *Discussion* should be thorough, but concise, and should avoid repeating information presented elsewhere in the document.

(30 – 40 % of essay.)

8. Conclusion

Summarize the key points from the discussion and reiterate how your investigation addresses the problem identified in the *Introduction*. Ensure that the conclusions you draw are supported by the information you present!

(≈ 5 % of essay.)

9. Acknowledgements

Acknowledge those who provided you with assistance. Acknowledge any organizations that funded the investigation.

Disclose any potential conflicts of interest.

Works cited *or* **Bibliography** *or* **References**

A list of the resources you used to prepare the essay. Most of the resources should be scholarly books and scholarly articles.

It is critical to double-check the citations; errors are surprisingly common.

– 3 –

… thesis template …

A clear, concise title that accurately reflects the content of the thesis and is understood by a broad scientific audience

by

<author>

in partial fulfillment for the requirements of <course or degree>

at

<institution>

<date>

... thesis template ...

A clear, concise title that accurately reflects the content of the thesis and is understood by a broad scientific audience

We accept this thesis as conforming to the required standard.

<Committee member 1>	<Committee member 2>
<Committee member 3>	<Committee member 4>
<Committee member 5>	<Committee member 6>

… thesis template …

Abstract

A concise overview and summary of the research question, research design, and key results. The abstract should inform readers why the research is interesting, important, and relevant; how the research fits with and expands our knowledge in this field; what research was conducted; and the key results of that research. The *Abstract* must itself bc a complete document as it is also published separately.

(The *Abstract* is on its own page, without page numbers. 100 – 300 words)

Table of contents

A table of contents that lists the chapters, headings, and numbered sub-headings and their page number. Appendices are included in the *Table of contents*.

Appendix B of *Communicating Science* explains how to use styles, which will dynamically create and update the *Table of contents*, *List of figures*, and *List of tables*. See, for example, the table of contents of *Communicating Science*.

(The *Table of contents* is on its own page(s).)

List of figures

A table that describes every figure, graph, and image, and their page numbers. It is reasonable to reproduce the figure captions in the *List of figures*.

(The *List of figures* page is optional. If used, it is on its own page(s).)

List of tables

A table that describes every table and their page numbers. It is reasonable to reproduce the table captions in the *List of tables*.

(The *List of tables* page is optional. If used, it is on its own page(s).)

The above sections either can be without page numbers or numbered using roman numerals. If using roman numerals, the title page is page i, the *Abstract* is on page iii, and the *Table of contents* on page v, etc. (assuming a double-sided document). However, do not show the page numbering on the title page.

1. Introduction

A review of the relevant scientific literature in this field and how your research — the subject of the thesis — draws together and builds upon the existing knowledge of this field. This section should cite numerous sources to illustrate the importance of your work and place your work in the context of what is already known.

– 1 –

... thesis template ...

The *Introduction* should answer four main questions: What is already known in this area? What research was conducted? Why was the current research conducted? What are the key results of the current research? Answering this last question foreshadows the results.

The goal of the *Introduction* is to convince the reader that the current research makes an important contribution to the field.

(The *Introduction* starts on page 1. 10 – 20 % of thesis.)

2. Materials and methods *or* **Procedure**

If your research used multiple similar methods, include the common aspects of the methods in this chapter and the unique details in the appropriate chapters.

If your research used many different methods, state this and move the *Materials and methods* information into the appropriate chapters.

For each experiment, give a detailed description of the equipment and supplies used to conduct the experiment. A schematic of the experimental setup aids in understanding. For equipment, include the make and model of the equipment, any modifications to the equipment, and the instrument settings during data collection. For supplies, include the manufacturer, lot number, and how the supplies were stored and handled prior to use in the experiment. Document all known hazards regarding the materials and procedures. Indicate the appropriate handling and safe use procedures.

Write this section as a process: each step in order. It must be sufficiently complete so that anyone with similar training is able to conduct the experiment. If the procedure was previously published, reference it but still give a detailed description and document any changes.

(10 – 20 % of thesis.)

3. <A short description of an experiment you conducted>

3.1 Introduction

A brief introduction to this experiment, focusing on the rationale for conducting this experiment and the results obtained.

3.2 Materials and methods

Any materials, instrumentation, or experimental procedures unique to this experiment not detailed in Chapter 2.

– 2 –

... thesis template ...

3.3 Results

Present the data obtained and the analysis of that data. Data is commonly presented in tables and figures, but the same data is not in both a table and a figure. Chemical and/or mathematical formulae explain how the results are obtained.

All the data collected should be included in the thesis. If there is a lot of data, include representative data in this chapter and all of the data in an Appendix. Data can also be included in electronic form on a CD or DVD, which is valuable if the experiment generated considerable electronic data, such as computational, modeling, and image/video data.

A well-formatted and comprehensive results section is critical as it forms the basis for showing the importance of your work.

3.4 Discussion

This section presents a detailed interpretation of the results in the context of the research questions and relates this new information to existing knowledge. The *Discussion* should be thorough, but concise; avoid repetition of information presented elsewhere in the document. Ensure that the statements you make are supported by the results you obtained!

(Each experimental chapter is 10 – 20 % of the thesis.)

4. ... <A short description of an experiment you conducted>

Repeat the steps in Chapter 3 for each experiment conducted.

8. Discussion (assuming chapters 3 – 7 describe different experiments)

This section draws together the individual discussion sections and applies the information to the overall research project. The *Discussion* should discuss the theoretical and practical applications of this new knowledge, and its significance to science and society.

The *Discussion* should be thorough, but concise; avoid repetition of information presented elsewhere in the document. Ensure that the statements you make are supported by the results you obtained!

(20 – 40 % of thesis.)

– 3 –

... *thesis template* ...

9. Conclusion

Summarize the key points from the *Results* and *Discussion* sections and reiterate how your research draws together and builds upon the existing knowledge in this area. Ensure that the conclusions you draw are supported by the information you present!

(≈ 5 % of thesis.)

10. Future work

Your research provides an incremental step toward better understanding this field of science. This section explains how your experiments could be improved to obtain better data and experiments the next person working in this field could conduct to further this work.

(< 5 % of thesis.)

11. Acknowledgements

Acknowledge those who provided you with assistance. This includes your research supervisor and everyone who assisted with any aspect of your research project: provided samples, equipment, scientific advice, ...

Acknowledge the organizations that funded the research.

Disclose any potential conflicts of interest.

Works cited *or* Bibliography *or* References

A list of the resources you cited when preparing the thesis. Most of the resources should be peer-reviewed scholarly books and scholarly articles.

It is critical to double-check the citations; errors are surprisingly common.

Appendices

Appendices contain all the experimental results and specialized information not critical to the discussion, but important to those conducting similar research. This includes derivations, unique mathematical calculations, and rationalizations for the procedures or analyses conducted. Ensure the thesis includes references to each appendix.

List appendices alphabetically: A, B, C, ...

– 4 –

5.7 Scholarly posters

A scholarly poster is a predominantly visual presentation of your research/investigative project. The limited space on a poster means you must *focus on one or a few key aspects* of your project.

Storyboarding a scholarly poster

Since posters are a predominantly visual document, storyboarding the information and layout on paper is especially important and will save you time when you begin creating the poster.

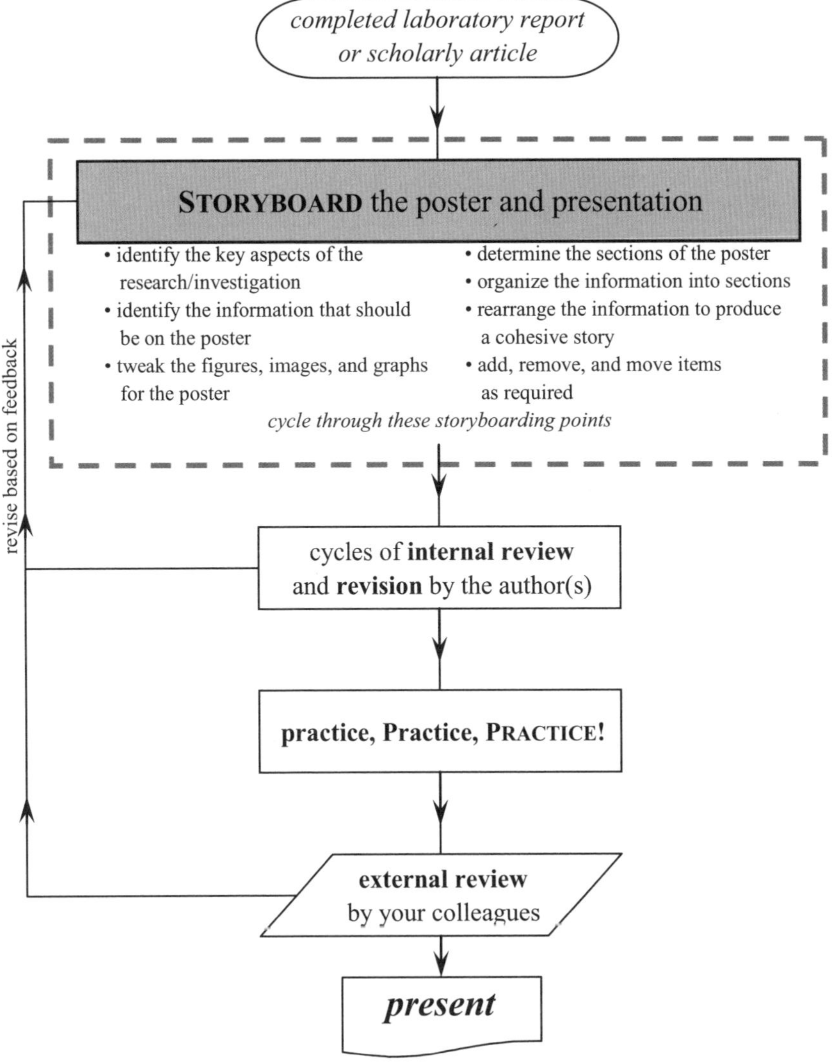

Figure 5.8 A flowchart outlining the steps to preparing a research or investigative poster.

A poster requires the project to be complete. You should have prepared either a complete laboratory report or essay on your project. This gives you a comprehensive understanding of the information, what it adds to the current understanding of the field, and its importance in society. You need this in order to

- understand the entire scope of the work
- identify and extract the key information for the poster
- draw upon the full understanding when presenting the poster
- answer questions on related topics

Layout of a scholarly poster

If your institution, department, or research group has a poster template, use it! Templates commonly have the institution logo and the fonts pre-configured. If you are starting from scratch, I recommend preparing your poster in Microsoft PowerPoint®, Adobe Illustrator®, CorelDraw®, or Microsoft Publisher®.

Posters are predominantly visual, with text supporting the figures, images, and graphs. Ensure the text occupies no more than 60 % of the poster area.

Posters typically start in the top left and conclude in the bottom right, consistent with how we read. Headings help guide the reader from section to section.

The poster session organizers will state the recommended and/or maximum poster size. Common sizes include

- 1.0 m × 1.5 m (40" × 60")
- 1.0 m × 1.2 m (40" × 48")
- 1.2 m × 1.0 m (48" × 40")

Figure 5.9 presents horizontal and vertical poster templates. At the poster session, you will see that actual posters deviate from these templates, which is fine! You must adapt the layout to suit your needs. By looking at other posters, you will get ideas to incorporate into your *next* poster.

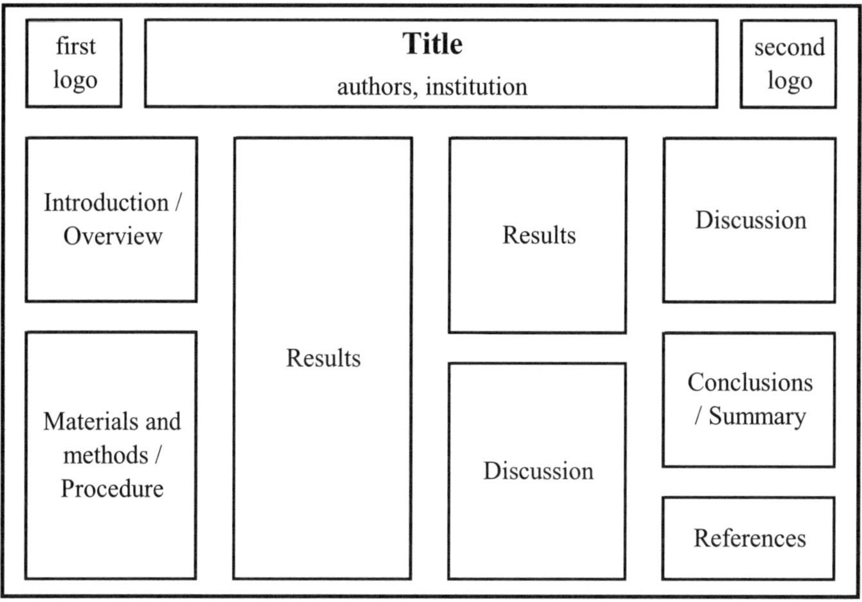

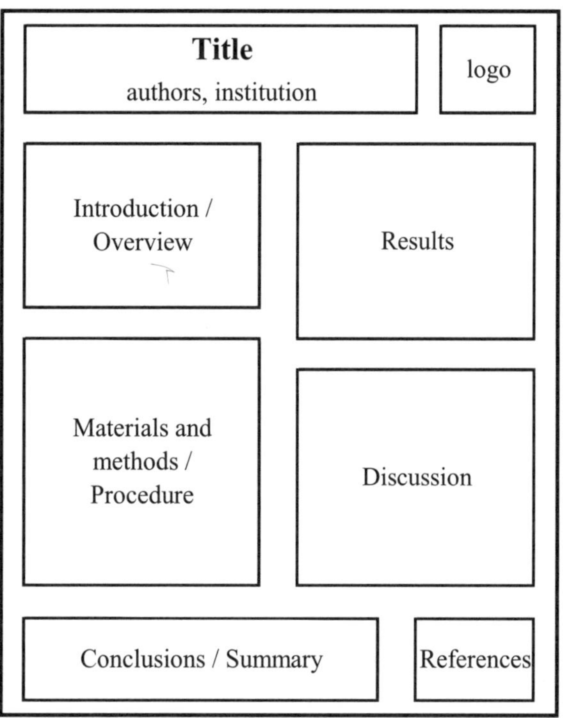

Figure 5.9 Sample templates for horizontal and vertical posters. Page 201 presents a selection of posters presented at scientific conferences.

Headings*

The templates in Figure 5.9 list numerous headings. The headings you use will depend on your discipline and the specific nature of your poster. In many cases, you will combine headings or use different headings.

Title: choose a title that fits on one line and is likely to attract visitors to your poster.

Author and Institution: include the first and last names of all of the authors and their institution(s). Underline the name(s) of the presenting author(s).

Logo: the institution's logo is always on the poster. The second picture could be the departmental or research group logo, or group picture.

Introduction or *Overview*: a short overview of the project. Keep it brief and interesting. Answer the following questions:

- Why was the research/investigation conducted (background)?
- What research was conducted (the research question)?
- What knowledge was gained from this project?

Include a picture if the picture helps to explain your motivation for conducting this project. (Aim for fewer than 200 words.)

Materials and methods or *Procedure*: a brief overview of the experimental design and equipment used. A schematic or flowchart is often easier to understand. (Aim for one or two figures and/or fewer than 200 words.)

Results and *Discussion*: these sections should be the bulk of your poster. Summarize the data in tables and figures. Use captions to explain each table and figure. The text should focus on interpreting and explaining the significance of the data. (Aim for up to five tables and figures and fewer than 400 words.)

Conclusion or *Summary*: summarize the project that was conducted and the key results. Use a bulleted list if appropriate. Explain the importance of the project to science and society. (Aim for fewer than 200 words.)

Bibliography or *References* or *Works cited*: list the references you cite in your poster. If you have more than ten references, list the most

* Adapted from

Purrington C. Advice on designing scientific posters [internet]. Swarthmore University; 2009 [cited 10 March 2010]. Updated version available from http://colinpurrington.com/tips/academic/posterdesign Used with permission.

important five to ten references on the poster and have all of the references on a separate sheet.

Other headings

Acknowledgements: list organizations and colleagues who assisted with funding, equipment, scientific advice, and/or logistical support. Funding bodies commonly have logos that you may use to acknowledge their support. Disclose any conflicts of interest in this section. (Aim for fewer than 40 words.)

Future work: given the results you are reporting in this poster, what are the next steps for the project?

Contact information: for visitors wanting more information, include your email address and the group website, if one exists. Ensure your poster, in PDF format, is available on the website.

The poster session

Posters are typically presented at a poster session, where tens to hundreds of people present posters at the same time. You have — at most — a few minutes to convey the key aspects of your work to visitors.[*]

Figure 5.10 A poster session typical of an academic conference.

Source: Lunar and Planetary Institute. Used with permission.

[*] Poster sessions are often held concurrently with social events. This sometimes makes it harder to get people interested in your work.

Preparing a scientific poster

Communicating Science

Roy H. Jensen, University of Alberta

Introduction

A scholarly poster is a predominantly visual presentation of your research/investigation. The limited space on a poster means you must focus on one aspect of your project.

This poster focuses on preparing a poster.

Fonts

- select one font for the title and headings and another font for the text (the same font can be used for both)
  - **sans-serif fonts** can be used for the title and headings
  - **serif fonts** are easier to read and better for the bulk of your text

Colors

Color figures are great. However, use color sparingly in text.

The focus must be your work, not on the colors and/or decorations in the poster. Posters with bright colors are hard to read.

Background: use one or two pastel colors or a faded background image related to your topic as a watermark to unify the poster. Try to integrate the background into the figures, images, and graphs.

Colorblindness: five percent of the population has some form of colorblindness. Red-green is most common.

Text

To be engaging

- write in the active voice
- present information in bullets and short sentences

The majority of visitors are not specialists in your field.

Use plain language (avoid jargon and acronyms) and focus on educating the *average* visitor. For the visitors who specialize in the same field, your verbal presentation will show your detailed understanding.

The text should compliment and explain the figures, images, and graphs.

- DO NOT USE ALL UPPERCASE (it's hard to read)
- use <u>underline</u>, *italics* , **bold**, and/or ***bold italics*** to emphasize text, but use them sparingly

Font size

- title: 100 – 120 pt
- authors and institution: 60 – 72 pt
- headings: 36 – 48 pt
- text: 24 – 28 pt
- captions and references: 18 – 24 pt

Title: should be easily readable from three meters (10 ft) away.

Text: should be easily readable from one meter (3 ft) away.

Figures, images, graphs: ensure the information can be clearly seen from one meter, including the caption text.

White space

White space is critical.

Ensure there is sufficient blank space surrounding text, tables, figures, images, and graphs so that the poster is easy to read and does not look crowded.

http://jfly.iam.u-tokyo.ac.jp/color/ provides suggestions on how to use color so that it is viewable by most people.

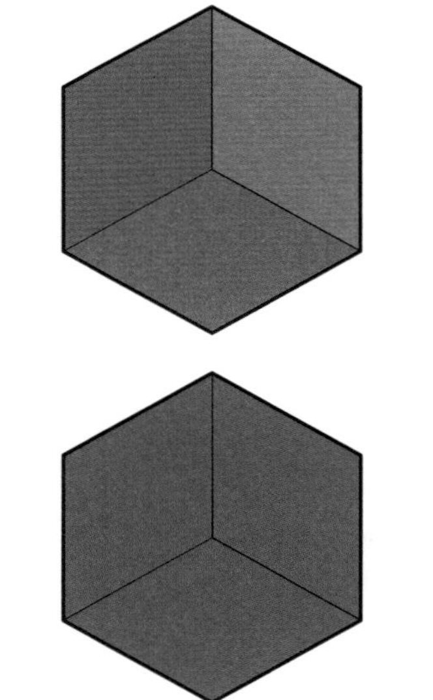

Figure 1. (top) Red, green, and blue sections with the same intensity. (bottom) Converted to greyscale, these colors are the same. Colorblind people experience this merging of colors, losing information in the image.

Additionally,

- use line spacing of 1.0 or 1.5 (1.0 is single spacing)
- use capitals sparingly (capitalize proper nouns and only the first word in the title and headings)
- aim for a total word count of fewer than 800 words

Printing your poster

If the poster is for a course or a science fair, individual pages can be printed on normal paper and glued onto a larger poster board. When preparing this type of poster, consider printing in landscape.

If the poster is for presentation at a conference, the poster should be printed as a single sheet using a wide-format printer. Laminate the poster if you are going to use it at several conferences.

Transport the poster in a poster tube.

Yes, this poster needs more images!

As a presenter, you do not have a captive audience. Your poster must grab the attention of the people at the poster session. Once you have attracted them to your poster, you must converse with the visitors in a manner that interests and engages them. *Your enthusiasm is important!* You want visitors to leave with a good impression of you and the work you have done. This means the project must be relevant, it must be presented in a logical and concise manner, and you must have a good understanding of the project and related scientific field. Visitors may provide feedback on what you have done, recommendations for alternate analyses, and suggestions on future steps for your project.

Presenting your poster

Stand beside your poster so the visitors can see the poster!

When a visitor approaches, smile, but do not launch into your presentation. Casual visitors will read either the *Introduction* or *Conclusion* or look at the figures. These parts of your poster must engage and interest the visitor. Once they have had 10 – 15 seconds to review your poster, you can ask them,

> Do you have any questions?
> Would you like to hear about my project?
> How much do you know about <your topic>?

Ask with enthusiasm!

Visitors will usually spend a maximum of two to three minutes viewing your poster. Asking, "how much they know" provides you with information on their current level of understanding. You can then tailor your presentation to build on their understanding. Your presentation could be a formal presentation or an interactive conversation, depending on the visitor. As a presentation, briefly explain why your project is important and focus on the key results of your project. Do not just read the text on your poster — the visitor can easily read that — your presentation must add information to what is already on the poster. Be prepared to start anywhere in your presentation as visitors may already be familiar with some aspects of your project and only want to know one or two things. A conversation is more casual and proceeds wherever you and the visitor have interests.

When presenting your poster, stand beside your poster (facing the visitor) or beside the visitor (facing the poster). Use an open hand to show which area of your poster you are presenting on. People commonly use their hands to aid in presenting their work. This is fine, provided you do not overuse them. Regularly make eye contact with your visitors to ensure they understand what you are presenting. If they appear confused, stop and begin a conversation that may better inform them about your work.

Section 5.10 provides strategies for giving presentations and answering questions effectively, and lists questions commonly asked by visitors.

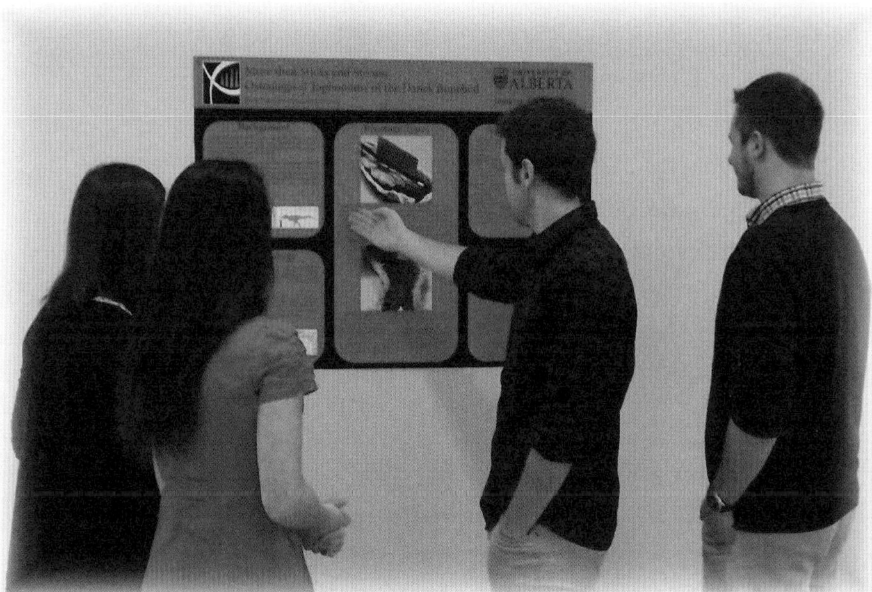

Figure 5.11 Ways to present a poster to visitors: standing beside the poster, facing the audience (top) and standing beside the visitor, facing the poster (bottom).

Reviewing posters

You want your poster to grab the attention of the visitor, be free of errors, be easy to understand, and be interesting. Keep the questions in Appendix A and below in mind when preparing your poster and presentation.*

Layout: this review should take about 60 seconds.

- Is the overall appearance inviting?
- Are the text and graphics aligned horizontally and vertically?
- Is there sufficient white space? Is it too crowded?
- Is there too much text? Too little text?
- Are there a reasonable number of figures, images, and graphs?
- Is the title catchy? Will it draw visitors?
- Is the title readable from three meters? Text from one meter?
- Is the flow of the poster obvious and easy to follow, *Introduction* to *Conclusion*?
- Are the key results easy to find?

Content: this review should take 10 to 30 minutes and the reviewer should read the entire poster a few times.

- Does the introduction grab the reader's attention and explain the project?
- Does the *Conclusion* convey the key points of the project?
- Is the poster focused and on-topic?
- Is the poster trying to convey too much information?
- Is the text written using bullets and/or short sentences?
- Does the text complement the figures, images, and graphs?
- Are the figures, images, and graphs focused on the topic?
- Are there any errors in the spelling or grammar? In the data or data analysis?
- Is the information presented at the level of the audience?

Presentation

- Is the presenter interacting with the visitor?
- Is the presenter adding information (not just reading the poster)?
- Is the information presented in a logical order?

* The layout and content questions are adapted from

Hess G, Tosney K, Liegel L. Creating Effective Poster Presentations. North Carolina State University; 2010 [cited 08 October 2011]. Available from: http://www.ncsu.edu/project/posters/

- Is the presentation interesting and engaging?
- Can the presenter start at various locations in their presentation?
- Is the presenter confident about the information?
- Does the presenter handle questions professionally?

Once you have prepared your poster and practiced* it a few times, have your colleagues review your poster and presentation. Give reviewers the Appendix A questions and the above questions and ask them for recommendations to improve the poster and presentation. Chapter 6 provides more information on the review process.

Additional suggestions

- Dress appropriately (business casual) and talk loudly enough for the visitors to hear you.
- Have business cards available to hand out during the poster session.
- Put your picture on the poster, in either the bottom left or right corner. This helps the visitors identify who the presenter is.
- Print mini versions of your poster printed on 8.5×11 or 8.5×14 paper. These can be handed out to persons who are interested in your work. (If prepared properly, your poster will be readable when shrunk to this size!)
- Print images in 3D and have 3D glasses available.
- If your poster is going to be unattended, use a mini audio recorder (like the ones found in greeting cards) to leave a message for visitors.
    - "Press here for a 60 second overview of my poster."
    - Audio recorders can also be useful for supplementary material. For example, a presentation on animal communication could have multiple buttons, each with a sample of a different call.
- The poster session is a great way to network. If you find yourself having a positive discussion with a visitor, consider asking them if they can recommend someone with whom you can continue your work or do similar work.

Sample posters

The following posters are illustrative of quality posters. They may not conform exactly to the above recommendations, which is fine! These recommendations are guidelines only. All posters are reproduced with permission.

* Read your presentation aloud as if you were presenting it. Make corrections as required to obtain a logical, coherent presentation.

The application of constraint theory to self-organization of naturally-occurring molecules

Benjamin D. Esham[1*], John C. Mauro[2], Roger J. Loucks[1], Garrett J. McGowan[3]
[1] Department of Physics and Astronomy, Alfred University [2] Science and Technology Division, Corning Incorporated
[3] Department of Chemistry, Alfred University * Corresponding author: bdesham@gmail.com

Introduction

Counting method

Example

Results

Our survey consisted of 63 naturally-occurring molecules, with an average of 12.8 constituent atoms. Of these, 54 were underconstrained, five were isostatic, and four were overconstrained.

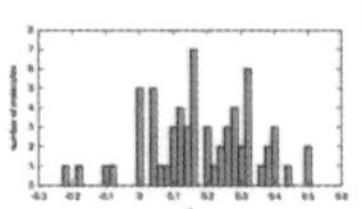

Figure 1. The values of f obtained in the survey

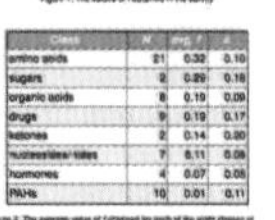

Class	N	avg. f	s
amino acids	21	0.32	0.10
sugars	2	0.29	0.16
organic acids	8	0.19	0.09
drugs	9	0.19	0.17
ketones	2	0.14	0.00
[illegible]	7	0.11	0.08
hormones	4	0.07	0.05
PAHs	10	0.01	0.11

Conclusions

As expected, the vast majority (86%) of molecules examined were underconstrained, with no molecules having an f value greater than 0.50. Of the four overconstrained molecules, three were polycyclic aromatic hydrocarbons (PAHs), fossil fuel constituents which take enormous energy to create.

Future work

Literature cited

Acknowledgments

For further information

Impacts of sea-level rise on Seattle, WA

DANIEL MAHR • BROWN UNIVERSITY • GEO132 • DECEMBER 12, 2009

Introduction

Data

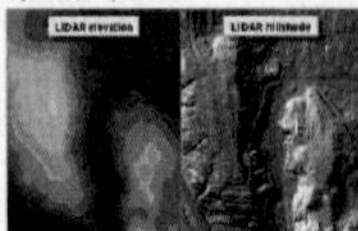

Methods

Results

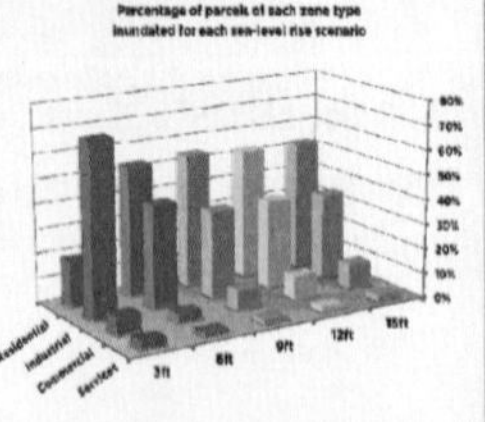

Discussion

Conclusions

Acknowledgements

Southern Flounder Exhibit Temperature-Dependent Sex Determination

J. Adam Luckenbach*, John Godwin and Russell Borski
Department of Zoology, Box 7617, North Carolina State University, Raleigh, NC 27695

Introduction

Southern flounder (*Paralichthys lethostigma*) support valuable fisheries and show great promise for aquaculture. Female flounder are known to grow faster and reach larger adult sizes than males. Therefore, information on sex determination that might increase the ratio of female flounder is important for aquaculture.

Objective

This study was conducted to determine whether southern flounder exhibit temperature-dependent sex determination (TSD), and if growth is affected by rearing temperature.

Methods

- Southern flounder broodstock were strip spawned to collect eggs and sperm for *in vitro* fertilization.
- Hatched larvae were weaned from a natural diet (rotifers/*Artemia*) to high protein pelleted feed and fed until satiation at least twice daily.
- Upon reaching a mean total length of 40 mm, the juvenile flounder were stocked at equal densities into one of three temperatures 18, 23, or 28°C for 245 days.
- Gonads were preserved and later sectioned at 2-6 microns.
- Sex-distinguishing markers were used to distinguish males (spermatogenesis) from females (oogenesis).

Histological Analysis

Male Differentiation

Female Differentiation

Temperature Affects Sex Determination

% Females

Temperature (°C)

(**P < 0.01 and ***P < 0.001 represent significant deviations from a 1:1 male:female sex ratio)

Rearing Temperature Affects Growth

Body Weight (g)

Temperature (°C)

Growth Does Not Differ by Sex

male

female

Body Weight (g)

Temperature (°C)

Results

- Sex was discernible in most fish greater than 120 mm long.
- High (28°C) temperature produced 4% females.
- Low (18°C) temperature produced 22% females.
- Mid-range (23°C) temperature produced 44% females.
- Fish raised at high or low temperatures showed reduced growth compared to those at the mid-range temperature.
- Up to 245 days, no differences in growth existed between sexes.

Conclusions

- These findings indicate that sex determination in southern flounder is temperature-sensitive and temperature has a profound effect on growth.
- A mid-range rearing temperature (23°C) appears to maximize the number of females and promote better growth in young southern flounder.
- Although adult females are known to grow larger than males, no difference in growth between sexes occurred in age-0 (< 1 year) southern flounder.

Acknowledgements

The authors acknowledge the Saltonstall-Kennedy Program of the National Marine Fisheries Service and the University of North Carolina Sea Grant College Program for funding this research. Special thanks to Lea Ware and Beth Shimps for help with the work.

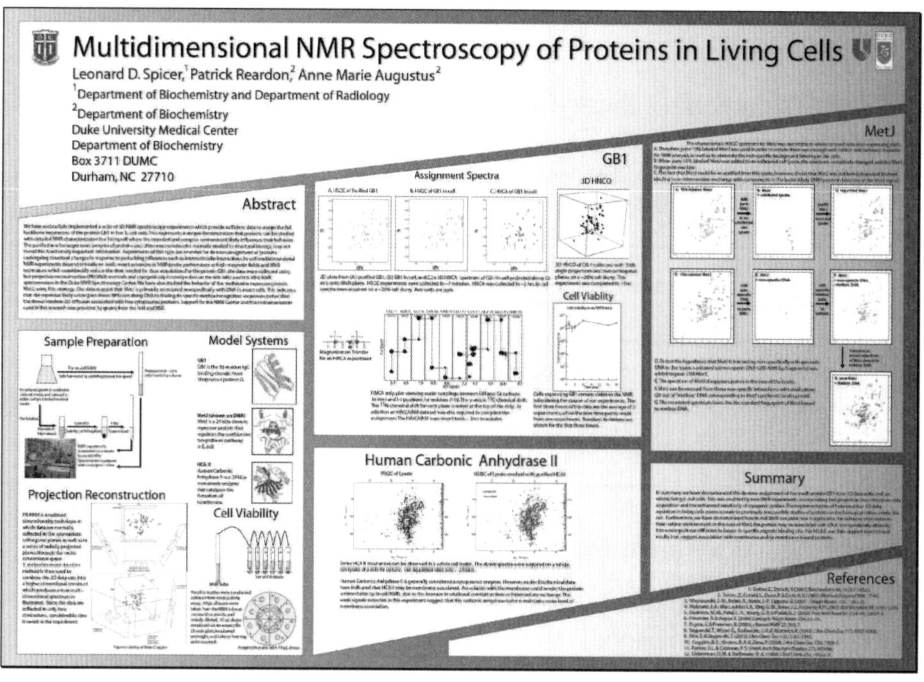

5.8 Preparing documents for other audiences

The style recommendations for the documents in sections 5.3 to 5.7 assume the readers are scientifically minded individuals. The documents have structures that scientists use and expect.

- *A document intended for a scientific audience* contains a logical process that leads from background → method → results → discussion → conclusion.

However, other readers do not have the same scientific background and have different interests in the scientific information, which must be considered when preparing documents for these readers.

- Administrators are often interested in the results and conclusions so they can make decisions. *A document intended for an administrative audience* often starts with the findings (results) and recommendations (based on the discussions and conclusions). The document then provides background and justification for the recommendations, with the methods and results often relegated to appendices. The *Executive summary* at the beginning of reports exemplifies this format.
- *A document intended for a public audience* often starts with a brief overview of the work and then goes into increasing detail for interested readers. This is seen in science magazines, newspapers, and encyclopedias.

Table 5.2 summarizes these differences.

Table 5.2 A comparison of documents prepared for scientific, administrative, and public audiences.

scientific	administrative	public
Documents contain a logical process that leads from background → … → conclusions.	Documents start with the findings and recommendations, and then provide background and justification.	Documents start with a brief overview of the work and then go into increasing detail for interested readers.

Preparing a document for another audience

Ask a scientist to explain their research to a non-scientist or to write an article for a newspaper, and you have asked them to do a very challenging task. Scientists often do not realize how little scientific knowledge non-scientists have or what misconceptions they may have regarding science. Furthermore, it is challenging to simplify advanced scientific theories, experiments, results, and implications into common language while still maintaining the accuracy of the research and without introducing misconceptions. Tables 1.8 and 1.9 (pages 41 and 43) list Latin and scientific terms and their plain-language equivalents. It may also be challenging to convince a scientist to prepare the document in a format expected by the audience, as illustrated in Table 5.2.

ScienceCartoonsPlus.com

It is critical for scientists to communicate their research to non-scientists. For politicians and administrators, an understanding of the relevant science allows them to make sound and rational decisions. For the public, awareness of the importance and applicability of science in everyday society is important so that the public supports and demands independent and unbiased scientific research. A scientifically literate public is less susceptible to false claims of scientific efficacy.

> Sadly, there are a growing number of disreputable organizations that capitalize on society's generally poor understanding of science. Advertising with false information tricks the public into purchasing products with zero or minimal benefits, and even purchasing products that can be dangerous.

Students are often best at preparing documents for other audiences because they bridge the gap between scientific and public understanding. Retaining this ability as you advance through your education and career will benefit you and science.

A primary goal when preparing documents for public audiences is to maintain scientific accuracy while translating high-level scientific terms and concepts into plain language.

ACADEMIC LANGUAGE: The linear integrity of the pipe was compromised.
The biota exhibited a 100 % mortality response.

COMMON LANGUAGE: The pipe was bent.
All of the plants died.

Storyboarding a document for public audiences

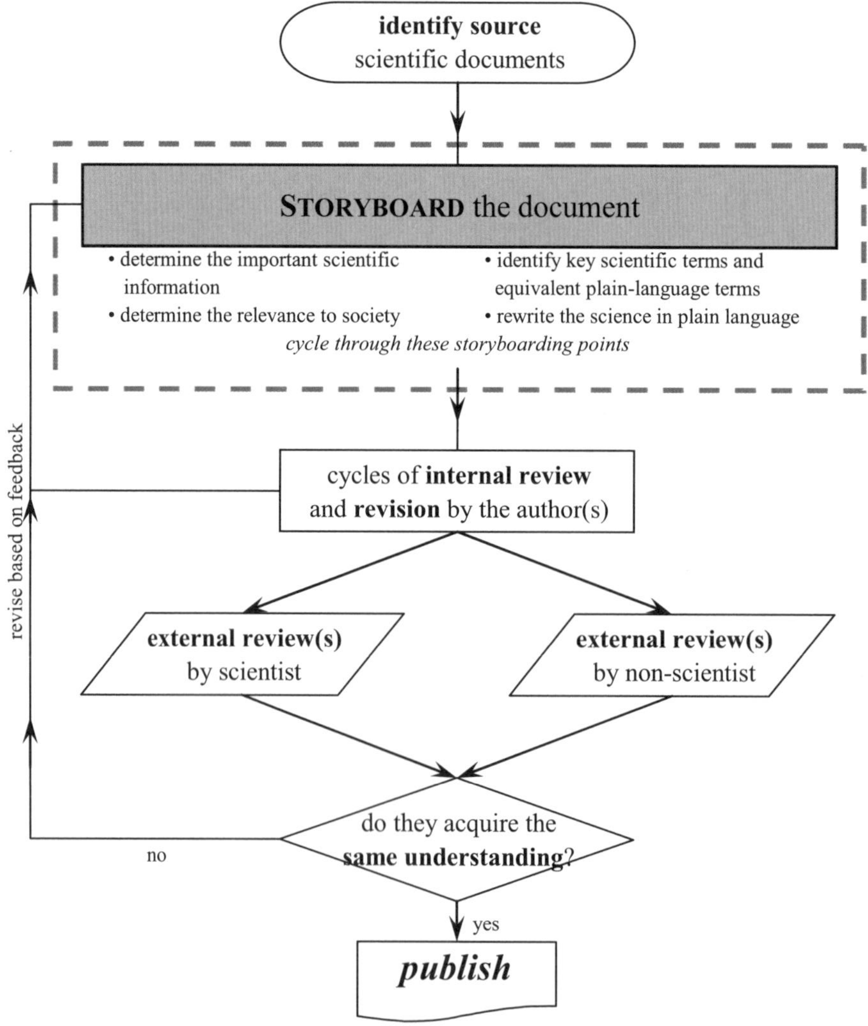

Figure 5.12 A flowchart outlining the steps to preparing a document for another audience.

Reviewing a document prepared for another audience

You want your document to be free of scientific and grammatical errors, easy to understand, and interesting. Ensure the readability statistics are appropriate for the audience (see Table 5.1). Appendix A lists questions you should keep in mind when preparing your document.

Once you have prepared your document and reviewed it once or twice,* have it reviewed by someone who understands the science and by someone who does not have a scientific background. Ensure they both reach the same, correct understanding of the science. Give reviewers the Appendix A questions and ask them for recommendations to improve the report. Chapter 6 provides more information on the review process.

> **"Global warming" and "Climate change"**
>
> Scientists have been investigating and documenting climate change for decades. There is overwhelming evidence that the Earth's climate is changing slowly, over decades, and that the change is primarily caused by humans. Scientific climate models predict changing global weather patterns (some areas warmer; some areas cooler), an overall increase in average global temperature, an increase in extreme weather, and many other climate changes.
>
> Instead of presenting all the technical data, climate scientists chose to focus on one consequence of climate change, *global warming*, because graphical data shows a gradual increase in average global temperatures since 1960. This increase correlates with increasing carbon dioxide concentrations. The scientists assumed that people would see the data, accept their conclusions, and advocate for change. However, climate change deniers used seasonal variations (cold winters) and localized cooler regions to seed doubt about global warming, despite these anomalies being predicted by climate models. The gradual increase over decades is difficult to detect on a day-to-day and season-to-season basis. Consequently, climate scientists have reverted to the general terms *climate change* and *extreme weather events* to illustrate the effects of climate change. Because extreme weather events occur regularly, are reported extensively by the media, and because their frequency is predicted to increase with climate change, these events and terms remind the public of the negative effects of climate change and encourage citizens to advocate for political change.

* Place the document aside for a day, or at least a few hours, and then reread it. You may be surprised that what made sense when you wrote it does not make sense now. Revise the document appropriately. (See page 131 and section 6.1 for details.)

5.9 Oral presentations

Oral presentations (colloquially called *talks*) occur frequently in science and in business. There are many types of presentations:

- *academic*: to inform others of your work and its importance
- *instruction*: to educate others and interest them in the information
- *proposal*: to obtain funding or permission
- *sales*: to encourage the audience to purchase what you are selling
- *self-promotion*: to a prospective employer or for promotion
- *team/group presentation*: to report on your individual progress to the other members of the team

As noted in Table 5.2 (page 204), the organization of your presentation depends on the intended audience.

A presentation requires a project to be complete. At minimum, you should have drafted and reviewed a complete laboratory report or essay on your project. This gives you a comprehensive understanding of the project, its contribution to the current understanding of the field, and its importance in society. You need this understanding so that

- you comprehend the entire scope of the work
- you can identify and extract the key information for the presentation
- you can draw on a full understanding when presenting
- you can answer questions on related topics

However, an oral presentation is not just a laboratory report or essay that you tweak into a presentation. An oral presentation presents information more like a poster than a written report. First, the limited presentation time means you must *focus on one or a few key aspects* of your project. Do not attempt to present everything you have done! Second, the audience only has one opportunity to receive the information, so you must present clearly and concisely with no extraneous information. You must choose information that the audience can reasonably understand at the pace you deliver it.

Storyboarding an oral presentation

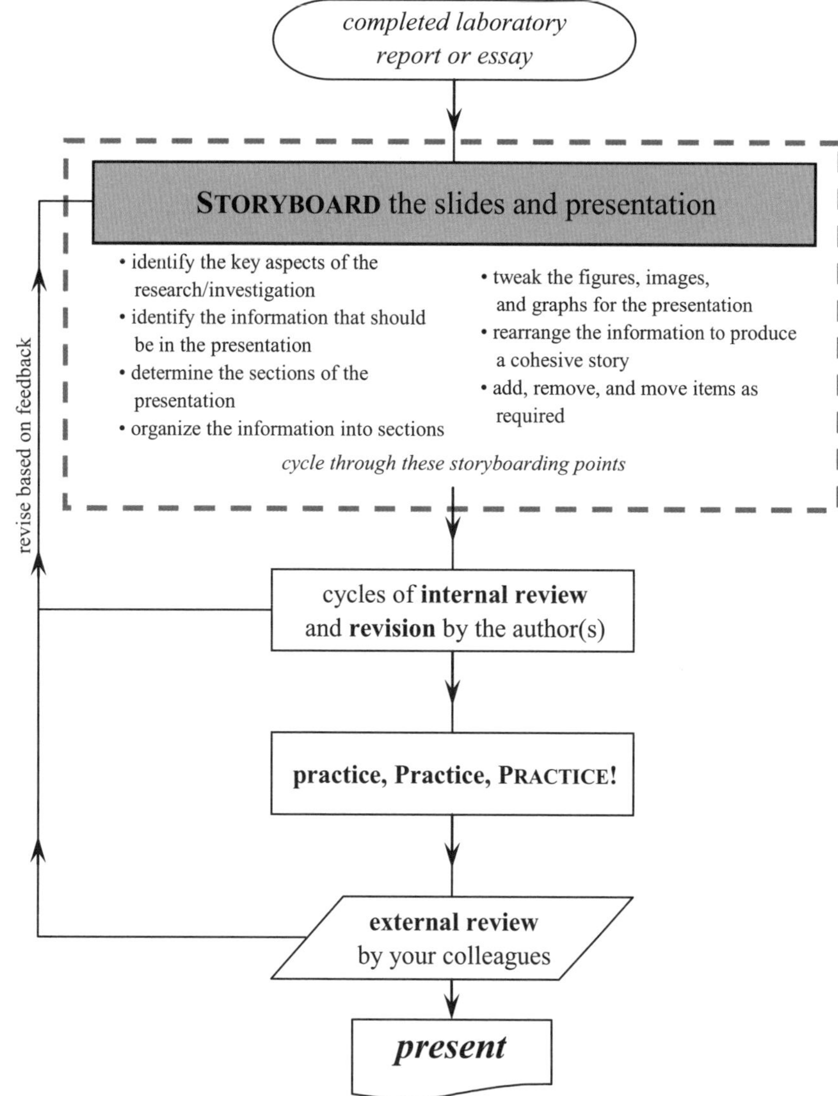

Figure 5.13 A flowchart outlining the steps to preparing an oral presentation.

Presentation software

Presentations are predominantly done using presentation software. Common software includes Microsoft PowerPoint®, Apple Keynote®, OpenOffice Impress™, and Corel Presentations™. Each program allows for dynamic presentations with color images, videos, animations, audio, and transitions. Appendix B provides guidelines for using presentation software.

> *Whatever program you use to prepare your presentation, remember that the audience wants to know what you have done and why it is important. The audience is not interested in your abilities to use the numerous features in the software. Only use the features of the presentation software that are necessary to best present the material. This means minimizing the number of animations in the presentation.*

Whatever technology you plan to use during your talk — data projector, overhead projector, whiteboard/blackboard/smartboard, flip chart, demonstrations, specialized software — make sure you are proficient in using that technology. Ideally, you should check the room a few days before your presentation to ensure the technology is available in the room and to visualize where to place the technology for maximum effect. Go to the back of the room and ensure the text and figures are visible and readable by those who will be sitting there. If time is available, practice giving the entire presentation in the room.

If you plan on writing during your presentation, make sure you write legibly and large enough so that your text is visible to the entire audience.

Planning your presentation

There are two common ways to organize your presentation:

- in the same format as a written report (formally, an ***exposition***)
- in the form of a story (formally, a ***narrative***)

For new presenters, the well-recognized and structured format of a written report — *Title*, *Introduction*, *Materials*, *Methods*, *Results*, *Discussion*, *Conclusion*, and *Acknowledgements* — is a useful guide to creating a presentation (see Section 5.3). It is reasonable to use the report-format as a safe format for new presenters. With increasing experience and comfort giving presentations, strive to increase the amount of storytelling in your presentations (see page 223). The headings you use can differ from those given below if they better explain the information.

Learn by watching others: what did you like about their slides and their presentation? What did you not like? Use this knowledge to improve your slides and your presentations.

Section 5.10 provides strategies for giving presentations and answering questions effectively.

Team presentations

Team presentations add a layer of complexity. The team members must agree on what each will prepare and say. The presentation must be prepared so that it is coherent and the transition between speakers is smooth.

Slides

If your institution, department, or research group has a slide template, use it! Templates commonly have the institution logo and the fonts pre-configured.

The amount of time you spend on each section depends on how long you have to present. Budget approximately two minutes for every slide. As you gain presentation experience, you may find that you spend more or less time per slide and can adjust the number of slides accordingly.

- 15 minute presentations: 7 – 8 slides
- 20 minute presentations: 9 – 11 slides
- 45 minute presentations: 20 – 25 slides

While the focus of all presentations is the *Results* and *Discussion*, you should have

Title: 1 slide with the presentation title and listing the first and last name of all of the author(s) and their institution(s). Underline the name(s) of the presenting author(s). Include the institutional logo and the date.

Outline: 0 – 1 slide listing the headings of your presentation. For long presentations where you are presenting on two or more topics, it is valuable to show the outline slide after each topic to remind the audience where you are at in your talk.

Introduction: 1 – 5 slides presenting a brief history of the research area, the context of your work, and the key findings. References should be at the bottom of each slide.

Methods: 0 – 3 slides giving a brief overview of the experimental design and equipment used. A schematic or flowchart is often easier to follow than text. Omit for investigative presentations and short research presentations. If omitted for a research presentation, briefly present the methods orally prior to giving the results.

Results: 1 – 6 slides presenting the key results of your project. Tables and figures are preferred.

Discussion: 1 – 6 slides explaining the significance of the results, placing the results in context of the other research in your field, and illustrating the importance of your project to society.

The *Results* and *Discussion* sections may be combined.

Conclusion: 1 slide summarizing the important aspects of the project.

Acknowledgements: 1 slide listing all those who assisted with the project (colleagues and funding organizations). Include your contact information for those people wanting more information.

Numbering the slides is at your discretion, but conveniently allows the audience to quickly identify which slide they are interested in.

You can also prepare slides that answer frequently asked questions: additional data, details on the data analysis, future work, etc. These slides are not shown during your presentation, but are shown if someone asks an appropriate question. Having these prepared shows that you have put a lot of thought and time into preparing your presentation.

Preparing slides for an oral presentation

Roy Jensen, M.Sc., Ph.D.

<date> <email address>

A presentation for inclusion in *Communicating Science.*

1

These slides have more text because they need to convey information without a presenter!

Outline

Slide templates

Fonts & style

Content

Engaging the audience

Review questions for presentations

Conclusions

2

Slide templates

Communicating Science

Templates

- choose a template and use it for all slides
  - presentation software has templates
  - you can create your own unique template
- simple templates are often better and easier to work with
- look for ideas in presentations given by other speakers
- to make an important point stand out, ***break the rules!***
  - use a different background, font, color, and/or animation.

3

Slide templates

Communicating Science

Colors

- two template strategies are common
  - ***dark text*** on a light background
  - ***light text*** on a dark background
- the background can be a gradient or pastel picture
- bad combinations
  - blue on black
  - yellow or pastel colors on white
  - red on green or blue

Dark background with a gradient fill.

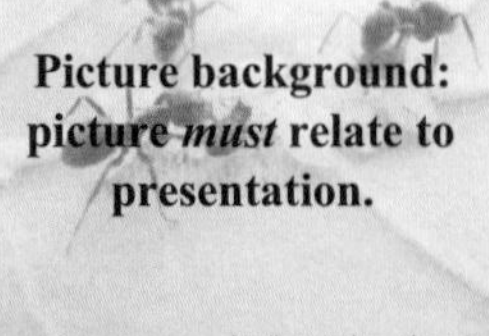

4

Slide templates

Communicating Science

Colors

- Five percent of the population has some form of color blindness.
  - Eight percent of men are red-green colorblind.
- http://jfly.iam.u-tokyo.ac.jp/color/ provides suggestions on how to alter images to make them viewable by most people.

Red, green, and blue sections with the same intensity.

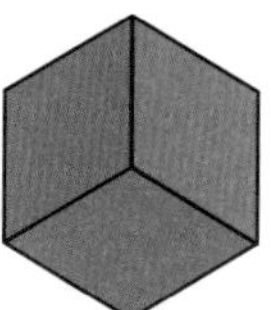

Converted to grayscale, these colors are the same. Colorblind people experience this merging of colors, losing the information in the image.

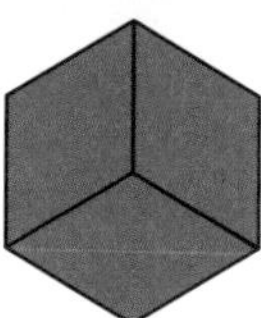

5

Fonts & style

Communicating Science

Font considerations

- use a single font for the title and headings, and another font for the text (the same font can be used for both)
- **sans-serif fonts** are easier to read on-screen
- **serif fonts** are easier to read in print
- DO NOT USE ALL UPPERCASE (it's hard to read)
- use underline, *italics*, **bold**, ***bold italics***, and **color** to emphasize text
  - use them sparingly to maximize their effectiveness
- ***ensure correct spelling and grammar***

6

Fonts & style

Communicating Science

Common font sizes

Title	**44 – 72 pt**
Heading	28 – 36 pt
General text	24 – 28 pt
General text may be indented	
➪ Bulleted subtext	24 – 28 pt
• bulleted subsubtext	20 – 24 pt
○ ditto	18 – 24 pt
1. Numbered subtext	24 – 28 pt
1. numbered subsubtext	20 – 24 pt
<large spacer>	24 – 28 pt
<small spacer>	10 – 12 pt

7

Content

Communicating Science

Slide layout

- ➪ present information in bullets and short sentences
  - *maximum*: two lines of text per bullet (ideally one line per bullet)
  - avoid long sentences
- ➪ use as few words as possible
  - expand on the words in your presentation
- ➪ use figures to present complex information
- ➪ ensure sufficient white space (blank space) on each slide
  - leave room around the edges and around figures
  - too little white space looks crowded and is hard to read
- ➪ ***the audience will not read paragraphs of text***
  - ***they will get annoyed if you read it to them***

8

Content

Communicating Science

Content

- decide on two to four key concepts that you will focus on in your presentation
  - with more time, more concepts can be presented
- no more than one concept per slide
  - budget two minutes per slide
  - it may take more than one slide to present a concept
- *focus on* ***simplicity*** *in your slides*
  - as few points as necessary to convey the information
  - figures are excellent for conveying information
  - figures and tables without captions are fine: your presentation explains the figures and tables

9

Content

Content

- only put essential information on the slides
  - ***minimize text***; ***MAXIMIZE figures***
  - your presentation adds detail
- write in the active voice
- use plain language
  - avoid jargon
  - define acronyms
- focus on educating the average visitor
  - to be retained, your material must integrate with their existing knowledge
  - specialists in the field will ask questions and/or contact you to discuss your research if they are interested

10

Content

Communicating Science

Animations and transitions

- software has numerous animations to *wow!* the audience

DO NOT USE THEM!

- too many animations
  - detract from your presentation
  - annoy the audience, sometimes to the point that they ignore you
- when watching other presentations, note how too many animations get annoying quickly
- if you must, use simple animations to display additional information and transition slides
  - **Appear** and **Fade** won't annoy your audience too much
  - *maximum*: three animations per slide (zero is a good number)

11

Content

Communicating Science

Figures (figures, images, and graphs)

- must focus on the information you wish to convey
- must be explained and emphasized to convey the importance of the figure to the audience
- are faster to comprehend than text
  - this facilitates understanding of complex concepts
- contain more information than text
  - specialists can get more information than you can convey verbally

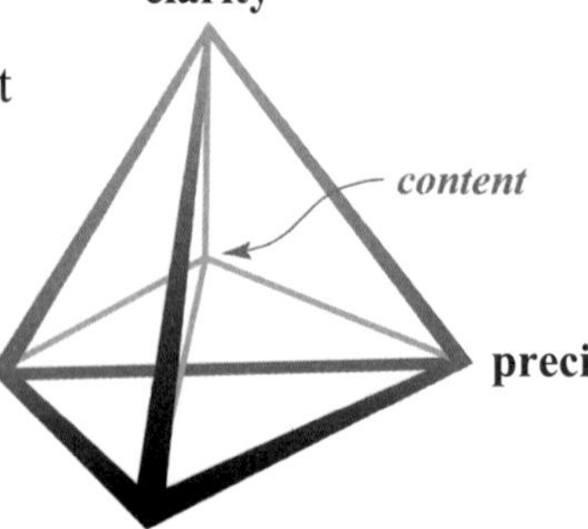

12

Content

Communicating Science

Tables

- ⇘ must focus on the information you wish to convey
- ⇘ if a report table has extraneous information, remake it to convey only the desired information
  - magnify the table so the font is readable
- ⇘ use a figure if you want to show trends in data

Document	Reading ease	Grade level
graduate documents; scholarly articles	< 30	> 14
undergraduate documents	30 – 40	12 – 14
popular science articles; technical reports	30 – 50	10 – 14
general audience (newspapers, websites)	50 – 70	6 – 10
children's books	> 70	< 6
Communicating Science	47	10

Content

Communicating Science

Multimedia

- ⇘ multimedia — images, audio, and video — add realism to the presentation
  - the audience quickly and more fully comprehends the information
  - audio and video clips should each be 15 – 60 seconds long
  - ensure the audio and video clips work on the presentation computer
- ⇘ explain what the audience should expect from the clip before playing it
- ⇘ all multimedia must pertain to the presentation
  - *clipart* — adding random images just to have images — *is bad*
    - random clipart detracts from the information you are trying to convey
  - cartoons are fine, provided they convey the intended message

Appendix B of *Communicating Science* provides details on how to add audio and video clips to your presentation.

14

Engaging the audience

Communicating Science

Negative aspects of presentation software

15

Engaging the audience

Communicating Science

Presentation

- your presentation is the *story* of your work
- your primary goal is to inform others of your work
  To do this, you must
  - catch and retain their interest
  - organize your presentation logically
  - focus the presentation to the audience's knowledge level
- additionally, you must
  - show enthusiasm for the material
  - turn nervousness into enthusiasm
  - stay within your time limit

16

Engaging the audience

Communicating Science

Presentation

- ***minimize text***; ***MAXIMIZE figures***
- your slides present key points
- your presentation expands on those points and develops the *story* of your research
- when you put up a slide, give the audience a few seconds to scan the slide
  - If the audience knows where you are going, they will pay attention to the *additional detail* you provide.

Handouts: some speakers distribute their presentation before they present. It gives the audience something to write on! But they can then skip ahead and read the conclusions before you get there.

17

Engaging the audience

Communicating Science

Laser pointers

To effectively use a laser pointer:

1. *turn it on*
2. use the laser to underline or circle a region to draw the audiences attention there
3. *turn it off*

Do not

- keep it on constantly
- use it for pseudo-animations

18

Review questions for presentations

Communicating Science

Review questions for presentations…

1. Is the layout appealing?
2. Is there a reasonable amount of text and images per slide?
3. Is there a minimum amount of animations and transitions?
4. Is the presenter facing the audience?
5. Is the presenter adding content (not just reading the slide)?
6. Does the presenter spend a reasonable amount of time per slide?
7. Is the material presented in a logical order?
8. Is the presentation interesting and engaging?
9. Is the presenter confident about the material?
10. Does the presentation fit in the allotted time?
11. Does the presenter handle questions professionally?

19

Conclusions

Communicating Science

Conclusions

- The content of your presentation
  - follows a template
  - can be read throughout the presentation room
  - is focused and on-topic
  - uses figures to convey information
  - uses simple animations, or no animations at all
- Your presentation
  - repeats the key findings of your work and the important implications thereof
  - adds information to the points on the slides
  - finishes confidently

20

Storytelling

The report-style presentation is a safe and effective *linear* presentation. The audience is led from point to point through the presentation. Linear presentations work with all audiences, but are they the most effective?

> Shawn Callahan, a methods designer with Anecdote.com, relates a story about an IBM engineer who gave a presentation on privacy at a technology conference. The engineer presented seven key aspects of privacy and related two stories pertaining to privacy issues. Callahan, the next speaker, asked the engineer for permission to conduct a little test with the audience. It turned out that no one in the audience could recall more than two of the seven key aspects of privacy without referring to their notes. However, many were able to recall the fine details of both stories. (Adapted from www.youtube.com/watch?v=0A-8TPx1X50. Used with permission. Retrieved April 2011.)
>
> Relating this to the constructivist learning model: data and facts have little context for readers to relate with. However, if built into a story that readers engage with, they more readily integrate the information into their understanding.

All cultures use stories to convey messages and morals. Good stories appeal to our emotions and have the power to persuade and motivate the audience. Stories are *non-linear*. Good storytellers, like novelists, weave together information in a way that engages the audience and draws them into the story. Some of the best storytellers are religious leaders. Some of the worst are scientists.

One problem with storytelling is that what you consider a good story is not everyone's idea of a good story. If the audience consists of persons who want to be there and who are interested in your presentation — conferences, political rallies — then you have the possibility of giving a successful non-linear presentation. However, new presenters and persons presenting to diverse audiences — classroom instruction, public audiences, on controversial topics — should initially style their presentation in the safer linear format. With experience and comfort, the amount of storytelling can increase to make the presentation more engaging and memorable.

If done well, audiences find story-format presentations more engaging and more memorable. However, presenting in the form of a story is more challenging. Realize that

- A poorly presented report-format presentation will still convey information, but will bore the audience.
- A poorly presented story-format presentation will leave the audience confused — no information will be conveyed.

When preparing a story-format presentation, you will probably spend more time per slide. Change your slides to

- budget three to five minutes per slide
- *really* minimize the amount of text
- increase the number of visuals in your presentation
- have each slide present a chapter in the story

Like a story, your presentation must weave together the information in a way that engages the audience and draws them in. One difference between the report-format and the story-format is the concept of *surprise*. In report-format, you inform the audience of the conclusions early and then spend the remainder of the presentation explaining how your research/investigation came to those conclusions. In story-format, you build suspense throughout your presentation and only near the end of the presentation give the conclusions and their significance.

Put simply, the report-format presentation is a starting point for creating presentations. Once you are comfortable presenting your work, increase the amount presented in story-format. The key phrase is *comfort*: you must be knowledgeable in the information and comfortable in front of an audience.*

Reviewing oral presentations

You want your presentation to be free of errors, easy to understand, and interesting. Keep the review questions in Appendix A and on page 222 in mind when preparing your presentation.

* Instructional environments require a different philosophy. Instructional presentations — where the audience is expected to learn the material — require information to be presented slower and more straightforward, similar to report-style presentations. Stories are still valuable in adding context to the information so that the information is better retained by students.

Once you have prepared your presentation and practiced* it a few times, present to your colleagues and have them provide feedback. Give reviewers the questions in Appendix A and on page 222 and ask them for recommendations to improve the slides and presentation. Chapter 6 provides more information on the review process.

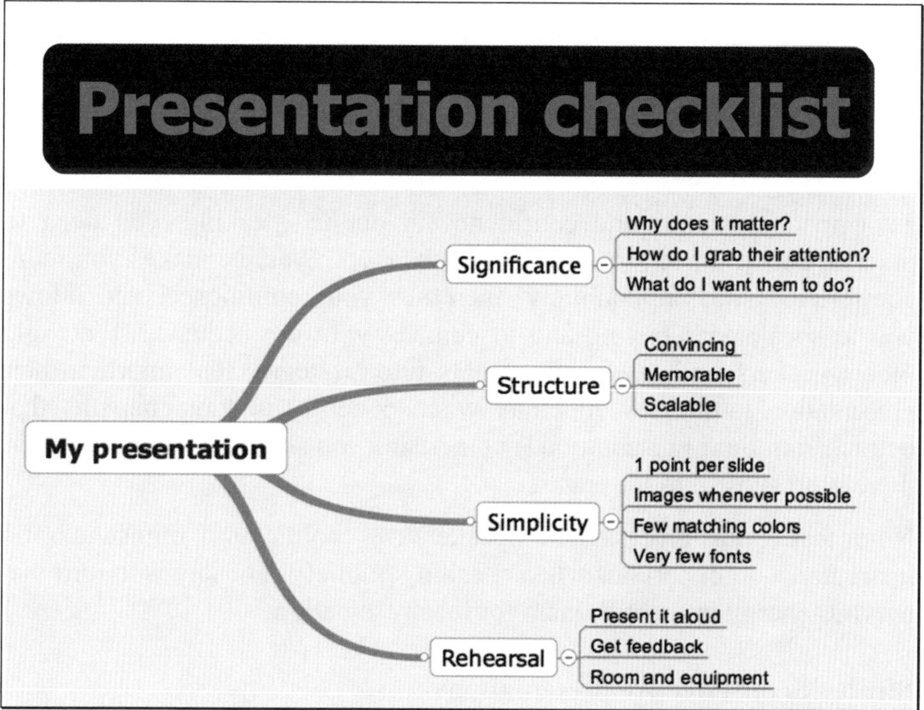

Figure 5.14 A visual representation of things to consider when preparing an oral presentation.†

* Read your presentation aloud as if you were presenting it. Make corrections as required to obtain a logical, coherent presentation.

† Source: Kapterev A. Death by PowerPoint (and how to fight it) [internet]. 2007 [cited 31 December 2011]. Available from http://www.slideshare.net/thecroaker/death-by-powerpoint/ Used with permission.

5.10 Public speaking

Studies show that people's greatest fear is public speaking! The fear of death is second, and the fear of spiders third.

You may not like speaking in public. However, if you wish to advance in your career, you will need to become a competent public speaker. The only way to do this is to speak publicly, and a classroom is the *safest* place to learn and hone these skills.

To be an effective speaker, you must engage the audience.

To gain competence and confidence in public speaking, you need to practice and you need to watch other speakers. Practice makes you more comfortable speaking publicly, increases your confidence, and allows you to speak with feeling and to connect with the audience. Watching other speakers will give you insights into strategies that improve their presentation (strategies you may wish to adopt) and mannerisms that detract from their presentation (mannerisms you should avoid). However, there is no substitute for practice.

Note that even professional speakers — politicians, ministers, and instructors — get nervous when speaking publicly. The key is to turn that nervous energy into enthusiasm for the presentation.

Strategies for effective presentations

Believe what you are saying. Sincerity is critical to credibility.

Prepare speaking notes. Your first set of speaking notes may be your entire presentation written out, but condense your speaking notes after every practice. Eventually, you should have only a few points for every slide, and those points remind you of everything you want to say. This will make your presentation more spontaneous and natural.

Smile when you speak. Smiling improves your mood and adds a friendly inflection to your tone. Always speak in a positive and professional tone.

Speak slowly, clearly, and loudly. People have a tendency to speak fast, quietly, and to mumble when they are nervous. Take a slow, deep breath before you start speaking and ensure you enunciate every word. Use simple language, the active voice, and short sentences. Realize that the audience only has one chance to understand what you are saying. Repetition of key points informs the reader what is important.

Treat the audience as a group of peers. Use words and language that you would use when talking to your professional peers. Colloquial language draws the audience into your work: "I did this …", "It was a surprise when …", "From this, we see that …."

Pauses add emphasis. Pauses give you time to breathe and the audience time to assimilate what you have stated. After showing a new slide, give the audience a few seconds to read the slide, and let them think about what you said and speculate on what you are going to say. (This engages them!) During the pause, maybe take a sip of water, but definitely take a deep breath before you start speaking again.

Add details and enthusiasm. If your slides are not too wordy, the audience will skim them to get an idea of where you are heading. You must then add context, details, and enthusiasm to what is on the slide. Your words and body language must express confidence and enthusiasm to draw the audience into your story.

Engage the audience. Design your presentation so that the audience gets involved in your presentation. Challenge them. Get them talking. Ask them questions. From their faces, you get real-time feedback on the level of engagement and can modify your presentation accordingly.

Figure 5.15 An instructor located to the side of their presentation and facing the audience. In this position, the instructor can look down at their notes, at the computer screen (on the instructors right), and can point to the projector screen without having to turn their back on the audience.

Use reasonable gestures. Non-verbal communication is critical to conveying enthusiasm and emphasizing key points. Hand and body motions, vocal inflections, and facial expressions are excellent non-verbal methods of doing this, but too many gestures distract from the presentation (see below). Realize that, in a large room, the people at the back will not be able to see your facial expressions, but they will be able to see your gestures. Make your gestures appropriate to your presentation.

Know and cater to your audience. What is the purpose of your presentation? Is the audience composed of scientists investigating similar phenomena? Are they a general scientific audience attending a departmental or conference presentation? Are they your fellow classmates or the public? What are the audience's expectations of you? The audience dictates how much detail you can present and how technical your language can be.

Talk to the audience. Do not talk to the screen. Adjust the computer monitor so that you can look down at the monitor and up at the audience without moving. Stand off to one side to avoid blocking the screen from the audience. Standing on the side allows you to use a pointer and still face the audience. Each time you speak, scan the audience or look at a different section of the audience. When speaking to one or a few people, such as during a poster presentation, make eye contact with them.

Keep to time. The host will usually inform you when you have two or five minutes remaining and will cut you off when your time is up. If you are running late, skip slides and move to the key sections: *Results* and *Discussion*. If you have to skip slides, it means you did not practice enough.

Practice, PRACTICE, PRACTICE! Practice your presentation five or more times by yourself and three to five times in front of various audiences before giving it to the real audience. Your academic colleagues should listen to your presentation at least once. Every audience should give you constructive feedback on your presentation, and you should revise your presentation based on their feedback.

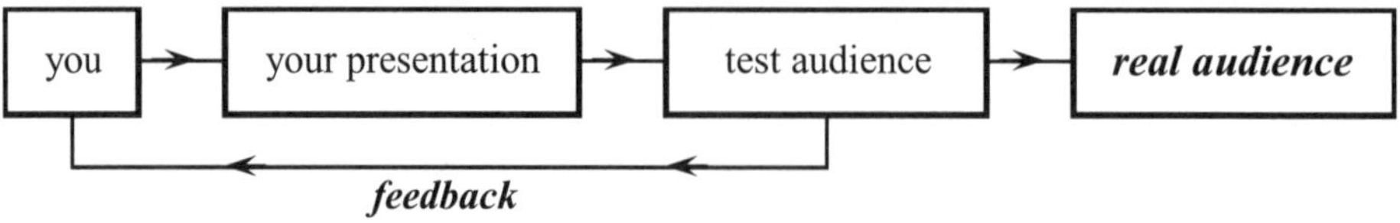

Figure 5.16 The process of preparing the best possible presentation to wow the audience.

If your presentation is too long, do not go faster, remove material!

Do not memorize your presentation and do not read off the slides. These presentations sound mechanical, lack emotion, and do not engage the audience.

Distracting mannerisms

As stated above, the occasional gesture is fine. Gestures are critical to conveying enthusiasm and emphasizing key points. However, repeating the same gesture throughout your presentation distracts the audience from your presentation. These distracting mannerisms become unconscious habits that you do without realizing and can alienate you from the audience.

Common distracting mannerisms include

- constantly clearing your throat or saying *um*, *ah*, or *like*
- speaking in a monotone voice
- constantly using your hands to emphasize statements (occasional hand gestures are fine)
- swaying from side to side or rocking forward and backwards
- gripping the lectern or tapping your fingers on the lectern
- repeatedly scratching a phantom itch or touching a part of your body
- constantly adjusting your clothes
- keeping your hands in your pockets, jingling keys, or coins
- constantly shuffling your speaking notes

Some hand motions may be perfectly fine in your culture, but are offensive in other cultures.

- pointing with the index finger, ☞ (considered aggressive and demeaning to many people worldwide)
    - if you must point with your hand, use an open hand, 🖐
- thumbs up, 👍 (offensive in most countries outside North America)
- the V sign, ✌ (offensive in most countries outside North America)

One way to detect distracting mannerisms is to videotape your practice presentations. (Yes, it is uncomfortable.) If you find something about your speech or mannerisms annoying, your audience probably will too.

Controlling nervousness

Your heart is racing. Your hands are sweaty. Your body is vibrating. You have butterflies in your stomach. You are speaking rapidly.

Adrenaline is a drug!

You have just taken a big hit of adrenaline. Whether it is your first presentation or your thousandth, you are *nervous*.

STOP! Think about any sport you enjoy and big moments in a game.

soccer	a free kick
hockey	a break-away
baseball	being at bat
swimming	diving from the high diving board
biking	going down a big hill
…	
going to school	taking an exam ☺

During these moments, you had adrenaline coursing through your body. How did you control it? Very likely, you panicked once or twice. However, you eventually learned to focus the adrenaline onto the task.

Public speaking is no different. Being nervous is normal. Use the following strategies to focus your nervous energy into giving an energetic and dynamic presentation.

- Be prepared to give the presentation — know the information and practice the presentation! This will give you confidence during your presentation.
- Before the presentation, engage in casual conversation with a few of the attendees.
- Take slow deep breaths before and during your presentation.
- Slow down: speak slowly and pause between slides.
- Look and act confident; speak with enthusiasm, force, and authority.
- Channel your nervousness into enthusiasm, expressed as an energetic and dynamic presentation. Focus on giving your presentation.
- Visualize yourself giving a successful presentation.
- Realize that you are there to convey information and that you are not yet a professional speaker. (The audience will accept that you are nervous.)
- Realize that the audience wants you to succeed. (Most of them have given presentations before and know what you are going through.)
- Do not expect perfection: if you make a mistake, ignore it or laugh at it. (Everyone makes mistakes and the audience knows what you are going through.)
- Accept assistance from the audience.

An important reason for learning and practicing presentations in school is that the environment is small and safe. These presentations give you experience and confidence. As you progress in school and your career, you will present on more challenging topics and in more challenging environments (like when your career depends on the presentation).

As you watch other speakers, you will see them become nervous and make errors. How do they correct for them? Are they successful?

Even with practice, you will find that you still become nervous before presentations. You may find that, once you start your presentation, you shift into *presentation mode* and channel your nervous energy into giving an energetic and dynamic presentation. You may even become excited about presenting because you like the feeling you have in presentation mode — akin to a runner's high.

If you suffer from extreme anxiety, consider joining a speakers club, like Toastmasters™. They assist individuals in overcoming their fear of public speaking. As you gain experience and comfort in Toastmasters, you will help new members in the club. The best way to learn is to teach!

Preparing for questions

You should be able to answer the following questions about your project:

- What research/investigation was conducted?
- Why is this research/investigation important?
- How was the research/investigation conducted?
- What did each of the authors contribute to the project?
- What are the key results? Why are they important?
- How was the data analyzed?
- What information do the figures provide?
- Are there any other interpretations of the data? Any unanswered questions?
- How does your research relate to the existing understanding in this field?
- What next? Where will you or someone else take this research?
- What are your particular plans for the short and long-term?
  (They may want to hire you!)

For investigative projects, you may also be asked:

- Why did you choose this topic?
- What resource was most useful?
- What was the most interesting aspect of your investigation?

Some of these questions may have been answered during your presentation. Your answer during the question period should provide more detail than what you gave during the presentation.

Anticipate questions the audience may ask and prepare answers. Include extra slides at the end of your presentation (possibly slides that you removed to shorten the presentation) to answer possible questions.

Answering questions

The following steps help you give the best possible answer and present yourself as a professional, attentive speaker.

1. Wait for the speaker to finish the question!
2. Repeat or rephrase the question. (This ensures the audience knows the question, focuses your mind on the question, and gives you a few seconds to formulate your answer.)
3. If you need to, take a few seconds to prepare your answer. It is better to speak succinctly than to ramble on.
4. Answer the question or state that you do not know the answer. (If you do not know — which is bad, but not as bad as fabricating an answer — ask for the speaker's contact information so that you can get the answer to them.)
5. Ask if you have answered their question.

5.11 Other forms of communication

The preceding sections detail how to prepare common scientific documents and presentations. However, much of your communication will be informal communication with your colleagues and collaborators. You may answer questions from clients and the public, and communicate on behalf of your organization. You will be communicating with people around the world — people whom you have never met and whose first language may not be English. It is important that you are clear, coherent, concise, and precise in your communication.

Memos, letters, and email

Memos, letters, and email are the most common forms of communication.

- *Memos* are commonly sent to persons within an organization (colleagues, management, superiors, subordinates).
- *Letters* are commonly sent to people outside an organization (clients, customers, suppliers, government).
- *Email* is supplanting memos and letters as a means of rapidly communicating with one or more people.

These documents are used to inform, update, ask and answer questions, and make announcements. They are typically short and focus on one or a few related topics. While these are common documents and often sent to people whom you know, they still must be professionally written, with proper spelling and grammar, and with language tailored to the recipient. Always write in a polite, positive, and professional tone, even if expressing concerns: your correspondence represents you and your organization.

Email is unique in that it is simple, convenient, and rapid. However, this does not mean that email should be informal or that unnecessary people should be copied on the email. A good practice is to compose your email, save it as a draft, and wait at least a few hours before reviewing, revising, and finally sending it. The more controversial the email, the longer you should wait and the more times you should review and revise it to ensure you are conveying the desired message.

Email is also used to submit other documents. In this latter situation, the body of the email is short, informing the recipient(s) of the attached document(s).

… memo template …

Corporate name logo

Date: <date>

To: <recipients>

From: <sender>

Re: <subject: a concise statement of the content of the memo>

The message should concisely present the reader with the issue and their required action (if any). Two types of memos are common:

- announcements
- issue notifications

Announcement memos

The first paragraph is the announcement.

The next paragraph(s) identify any consequences of the announcement.

The last paragraph indicates to whom questions should be directed.

Issue notification memos

The first paragraph presents the issue and the desired action.

The next paragraph(s) provide details on the issue so that an informed decision can be made. Each paragraph should present different aspects of the issue. This information can be presented in point form. Detailed information (reports, budgets, timelines, etc.) may be attached to the memo.

The last paragraph summarizes the details, repeats the desired action, and explains why this action is recommended.

Write in a polite, positive, and professional tone, even when expressing concerns.

Memos should be fewer than two pages long, preferably less than one page.

... letter template ...

ROY JENSEN

www.consol.ca
Roy.Jensen@consol.ca

119 Glencoe Boulevard
Sherwood Park, AB T8A 5J5
<date>

<address of recipient>
<address of recipient>

Attn: <recipient>

Re: <subject>

Letters are more formal than memos. I have taken the liberty of modeling this letter on my personal letterhead.

Letters have several sections:

- *letterhead* at the top contains the company name (or your name), logo, and contact information, including mailing address
- *address* contains the full mailing address of the recipient
- *salutation* identifies the recipient using their full name and title
- *subject* concisely states the content of the letter so that the recipient knows what your letter is about
- *body* presents the reader with the information you wish to convey. The format can be similar to the memo template.
- *closing and signature* personalizes the letter
- *enclosures* identify material attached to the letter
- *cc* identifies other recipients

There are two common letter layouts: block and modified block.

- *block layout* has all information left-aligned
- *modified block layout* has the date and closing to the right of center

This letter template is in modified block format.

Write in a polite, positive, and professional tone, even when expressing concerns.

Letters should be fewer than two pages long, preferably less than one page.

Roy Jensen, M.Sc., Ph.D.

Enclosures: <list the enclosures>

cc: <list the persons copied on this letter; cc stands for 'carbon copy'>

```
Date: Tue, 8 Feb 2014, 13:46
From: Aiden Campbell <A.Campbell@McGill.ca>
To: Dr. Kelly Fischer <Kelly.Fischer@McGill.ca>
Subject: Research assistant position

Dr. Fischer

You may recall our conversation in your office regarding the
research assistant position in your laboratory this summer.
Attached is an application consisting of a cover letter, my
curriculum vitae, and my academic transcripts.

The opportunity to work in your laboratory will provide me
with valuable experience and I am confident I will be an
asset to your research program.

Thank you,
Aiden Campbell

Fourth-year geology honors student
McGill University
```

Email software can automatically add electronic signatures. Use signatures to provide the reader with information about you that is not otherwise conveyed in the email. Signatures should be concise and well considered — they provide the reader with information about you.

Professional signatures should contain your name, position, and contact information.*

```
 Dr. Roy Jensen
(≡≡≡≡≡≡≡≡≡≡≡≡≡≡)————————————————————————————————————¤
 Lecturer, Chemistry
 E5-33F, University of Alberta
 780.248.1808
```

Personal signatures may contain a catchy phrase, statistic, or picture. Be careful designing your personal signature since you may inadvertently send an business email from your personal account. You don't want the signature to embarrass you or your company.

```
Roy Jensen
   If I have seen farther than others, it is because I was
   standing on the shoulders of giants. (Sir Isaac Newton)
```

* The ASCII art represents a laser and refers to my background in physical chemistry.

Instructions and procedures

Section 5.8 explored strategies for conveying scientific information to public audiences. This section goes one step further: it addresses how to write instructions for public audiences so that they can complete a technical task safely. These tasks include determining the chlorine content of hot-tub water, handling corrosive cleaners, using and servicing power tools (saws, lathes, lawnmowers), and installing memory into a computer. An error in the instructions could cause injury or death, or destroy expensive equipment. You must write instructions that are clear, coherent, concise, precise, and at the level of the reader.

Instructions are detailed step-by-step directions on how to complete a task. Instructions are written for people with little or no technical knowledge.

Procedures detail the steps required to complete a task. Procedures are written for people with some technical knowledge and may omit fine details that a skilled person would know.

> Laboratory manuals contain *procedures* for conducting the experiments and may contain *instructions* for operating specialized equipment.

Both instructions and procedures must be easily read and understood by the intended audience and must provide sufficient information so that the steps can be completely safely. When preparing them, complete and document every step, and repeat the process several times to ensure your documentation is complete. A video of the process is helpful so that you can review the video as you write the instructions or procedures. The following guidelines will help you prepare quality instructions and procedures:

- Write in the present tense.
- Keep the steps as brief as possible. If long, break into multiple steps.
- Pictures or illustrations may aid in visualizing some steps.

When writing *instructions*, you are assuming the reader has little or no technical skills. First, give an overview of the task to be completed. List the knowledge and skills the reader is expected to have, what tools and space are required, and the typical completion time. Second, list the hazards the reader may encounter and identify the symbols used to identify dangers, warnings, cautions, and notices in the steps.*

* Adapted from Wikipedia. Precautionary statement (internet). Accessed 17 June 2013. Available from http://en.wikipedia.org/wiki/Precautionary_statement

Danger: advises of an imminent hazard that will result in severe injury or death.
Warning: advises of potentially hazardous situations, which, if not avoided, could result in serious injury or death.
Caution: advises of potentially hazardous situations, which, if not avoided, could result in minor personal injury.
Notice: advises of hazards that could result in non-personal injury, such as product or property damage.

Figure 5.17 Color-coding of safety symbols used to identify dangers, warnings, cautions, and notices in instructions.

Finally, detail each step. In each step, identify the purpose of the step, tools/equipment required, relevant hazards, and what the reader must do to safely complete the step. Each step must be numbered. A challenge when preparing instructions is to keep them clear and concise because many details are required.

When writing a *procedure*, you need to determine what technical skills the reader has; this determines the amount of detail required.

A procedure that involves a distillation would have several steps in a first-year laboratory manual, but would only require one-step in a fourth-year manual.

Often preceding a procedure is a description of the project and a list of equipment and supplies that are required. Hazards associated with any of the equipment and supplies are also listed. In the procedure, list the major tasks necessary to complete the project. A flowchart may be included as a visual overview of the procedure, such as the flowchart in Figure 2.1 (page 92). Each step must be numbered. For some steps in the list, you may need to include the required equipment or more detail on how to complete the step, depending on the skill level of the reader.

Once you have drafted your instructions or procedure, persons with the appropriate knowledge must test them to ensure they are clear, coherent, concise, and precise.

Meetings, phone calls, and video calls

Chapter 1 explains that communication is 10 % verbal or written and 90 % non-verbal (tone, demeanor, dress, etc.). Written communication must be exact because the recipient does not receive the non-verbal cues. However, you convey non-verbal information during interactive events, from phone calls to in-person meetings. Humans are adept at noticing non-verbal cues: your clothes, posture, tone, eye movements, and a myriad of other cues inform the recipient of your interest, enthusiasm, and sincerity. Review sections 5.9 and 5.10 for strategies on how to effectively communicate orally.

Preparedness is important. Be prepared for meetings and scheduled calls by having read and printed the agenda and attachments, and by bringing a means of taking notes (pen and paper, computer, etc.). At the meeting, pay attention and involve yourself in the discussion when appropriate. Speak in a positive and professional tone, and stay focused on topic.

In addition to the events discussed above, you need to be aware of your non-verbal communication when chatting with someone in the hallway, when asking a question of a speaker at a conference, and when interviewing for a job. The better you are at communicating, the better your intended message will be conveyed.

Résumés and curricula vitae

A ***résumé*** is a one-to-two page document given to prospective employers that highlights your education, experience, and qualifications for the position. These are listed in reverse-chronological order (most recent listed first), with the most relevant section listed first. For example, if you just completed school and are looking for a job, order the sections: *Education*, *Qualifications*, *Experience*. If you have been working for many years, order the sections: *Experience*, *Qualifications*, *Education*. Customize each résumé to the employer and to the specific position you are applying for.

A ***curriculum vitae*** (CV) is an academic résumé, provided when applying for teaching and research positions. A CV summarizes all the work you have done in your academic career: education, academic positions, research experience, teaching experience, publications, persons you supervised, and funding received to conduct research. There is no length limit on CVs, and it is common for CVs to be tens of pages in length.

Some people and institutions separate their teaching accomplishments from their research accomplishments. While a CV focuses on research, a ***teaching dossier*** summarizes your teaching philosophy, teaching experience, and contributions to pedagogy.

Templates for résumés, CVs, and teaching dossiers are available on the internet. Academic institutions may have CV and teaching dossier templates. Also on the internet are the résumés and CVs of thousands of people. Use these to guide the creation of your résumé, CV, and/or teaching dossier.

Spelling and grammatical errors in these documents suggest many things — none of them positive — and decreases your chance of being hired. Most institutions have a career center that will review your résumé and/or CV and provide feedback on content and formatting. Also, consider asking instructors in the discipline and people already working in the field to provide suggestions and review your résumé and/or CV.

All job applications should start with a ***cover letter*** that introduces you and personalizes your application to the prospective employer. It is the first impression the prospective employer has of you. Your cover letter should

- indicate your familiarity with the company, the work that they do, and the specific work that you will be doing
- summarize the relevant education, experience, and qualifications that you bring to the company
- explain how hiring you is beneficial to the company
- finish by restating your interest in the company and in meeting with them to discuss your application and what you bring to the company

Social media: public forums and blogs

Correspondence posted to online forums (Facebook, Twitter, discussion groups) and posted to your personal weblog (blog) is public. Everyone can read it: your colleagues, employer, future employer, and those intentionally wanting to pick apart your arguments (especially if you are posting on a controversial topic). What and how you post informs readers about your personality and how you interact with others. Posting well-reasoned, intelligent statements shows readers that you are competent and willing to engage in academic discussion. Posting extravagant statements or making demeaning or harassing statements says something else.

To illustrate the influence of social media, search for your name on the internet. How many people have the same name? Where are you in the list? For the hits that do represent you, what impression would another person, such as a prospective employer or your colleagues, form of you? Is that impression accurate? Do you like the public *you*?

Unless you are the designated spokesperson for your employer, postings are your own opinion. However, your postings reflect your personality, your attitude, and possibly your employer. Because of this, you must be extra careful about what you post and how you communicate it. All of your posts — personal and professional — must be written formally and professionally. Reviewing your post before submitting it is critical to conveying the desired message.

Do not post in anger! If you let your anger or frustration spill out, you will most certainly live to regret it. If you are posting on a controversial topic or disagree with what someone else has posted, wait. Wait. WAIT several hours to a few days before posting. Review what the other person has posted to ensure you have understood them correctly. Have a colleague review all the documents before you post your comments. [Who am I kidding? You'll probably learn the hard way, just like I did.]

Media inquiries

The media serves an important function by informing the public about current events. In the context of *Communicating Science*, you will likely be interviewed on your academic accomplishments. However, before answering questions from the media, consider:

1. Have you read the institutional media relations policy?
2. Are you permitted to talk to the media?
3. Are you the most appropriate person to interview about this topic?

Even if a reporter is asking questions, you must not answer if you are unsure about whether you are allowed. However, saying "no comment" suggests you have something to hide. The best response is to say, "I am not authorized to answer questions." or "Please provide me with your questions in writing and <our organization> will respond promptly." The latter strategy — asking for the questions — allows you to consider each question and provide a clear, coherent, concise, and precise response.

If your research is such that you want to share it with the media (a news release) or that the media may be interested in it, it is worthwhile identifying the key aspects of your research you wish to convey and repeating them during the interview(s). Respond promptly to all media inquiries because reporters are working on deadlines, and you probably want public recognition for your work.

When approached by a reporter, ask them to explain what story they are working on. This will allow you to tailor your answers to the story. In your verbal or written response, the following strategies will protect you and provide accurate information to the media.

- Be honest. Keep your answers brief and to the point.
- Be positive and professional. Reporters form an opinion of you based on their brief interaction with you. This opinion influences how they write their article.
- Speak only on topics that you are knowledgeable about. Stick to the key aspects of your research; do not guess or speculate on topics that you are not familiar with. Do not let the interview go off-topic.
- Speak in a positive and professional manner, even when dealing with a difficult situation.
- Remember that when speaking to a reporter, you are speaking to the public. Be brief; reporters are looking for short sound bytes (typically 10 to 20 seconds long). Use common terms and communicate at the level of the public.
- Assume that everything you say is "on the record". Stating something "off the record" gives reporters a direction for future investigation, so the information will eventually become public.
- If you are unsure about how to answer or what information you can provide, state so. Then, after the interview, prepare an answer and promptly submit it to the reporter.

Everyone wants to ensure the correct information is communicated to the public. You may want to review the story before it goes to the public. However, most journalists do not permit this because of deadlines and because such a request questions the journalists' professionalism. To ensure an accurate media report, ensure that correct information is communicated during the interview. If the media report does contain minor errors, it is best to ignore them. If the report contains major errors, contact the reporter and positively and professionally inform them of the errors.

Finally, if you have been interviewed, inform others that should know, including any funding organizations, so they are aware and can be prepared for when the story is published.

Additional resources ...

... on the process of science, including the scientific method, research methodologies, and scientific communication

Process of Science [internet]. Vision Learning. [cited 18 October 2010] Available from http://www.visionlearning.com/library/cat_view.php?cid=49

... on writing a research article

Fischer B, Zigmond, M. Components of a research article [internet]. University of Pittsburgh; (undated) [cited 05 July 2012]. Available from: <no longer available>

Fischer B, Zigmond, M. Twenty steps to writing a research article [internet]. University of Pittsburgh; (undated) [cited 05 July 2012]. Available from: <no longer available>

AIP Style Manual (internet). 4th ed. New York: American Institute of Physics; 1990. Available from http://www.aip.org/pubservs/style/4thed/AIP_Style_4thed.pdf

... on preparing scholarly posters

Hess G, Tosney K, Liegel, L. Creating effective posters and poster presentations [internet]. North Carolina State University [cited 10 March 2010]. Available from http://www.ncsu.edu/project/posters

Kapterev A. Death by PowerPoint (and how to fight it) [internet]. 2007 [cited 31 December 2011]. Available from http://www.slideshare.net/thecroaker/death-by-powerpoint/

Purrington C. Designing conference posters [internet]. Swarthmore University; 2009 [cited 10 March 2010]. <no longer available> An updated site is available at http://colinpurrington.com/tips/academic/posterdesign

... showing sample scholarly posters

F1000 Posters [internet]. Available from http://f1000.com/posters

PhD Posters. Gallery of Work [internet]. Available from http://phdposters.com/gallery.php

Hess G, Tosney K, Liegel L. Examples of Posters [internet]. Available from http://www.ncsu.edu/project/posters/ExamplePosters.html

... on preparing effective presentations

Hansen K. A Dozen Eye-Openers about Telling Stories in Presentations [internet]. 2009 [cited 18 March 2010]. Available from http://astoriedcareer.com/2009/08/a-dozen-eyeopeners-about-telli.html

Gallo C. The Presentation Secrets of Steve Jobs [internet]. Available from http://www.slideshare.net/prwalker/the-presentation-secrets-of-steve-jobs-2814996

... on controlling nervousness

Controlling nervousness [internet]. Centre for Teaching Excellence. University of Waterloo; (undated) [cited 18 August 2011]. Available from: https://uwaterloo.ca/centre-for-teaching-excellence/teaching-resources/teaching-tips/lecturing-and-presenting/delivery/controlling-nervousness

Chapter 6. Peer review and peer evaluation

In your courses, you have learned the course material: biology, chemistry, earth science, mathematics, physics, English, etc. In addition, you were unconsciously evaluating the quality of the instructor, instruction, and course material. Did you like how the instructor taught? Was the textbook valuable to learning? You may have provided formal student feedback at the end of the course.

In Section 5.4, you were expected to assess the quality of scientific information in scholarly articles. In this chapter, you will apply these assessment skills to help your peers prepare quality documents and presentations. While this may feel uncomfortable at first, a classroom is the *safest* place to learn and hone these peer review skills, and these skills are critical for all professionals.

In your courses, you have likely experienced instructors deducting marks for spelling, missing or incorrect words, run-on sentences, and prose that does not make sense. Correcting these errors before submitting the document to your instructor would have improved your grade.

In your career, you will be required to prepare reports and give presentations to team members, management, clients, government officials, and the public. You want these documents and presentations to be of the highest quality; they cannot have errors. Poor-quality documents and presentations will negatively affect your company and could mean losing a contract or you losing your job.

> Roger Communications Inc. and Aliant Inc. signed a contract whereby Aliant would string Roger's cable lines. The contract states
>
> > This agreement shall be effective from the date it is made and shall continue in force for a period of five (5) years from the date it is made, and thereafter for successive five (5) year terms, unless and until terminated by one year prior notice in writing by either party.
>
> Rogers assumed it had a five-year fixed-rate contract that would be renewed every five years unless a party gave twelve months' notice. Aliant assumed the contract could be cancelled by giving twelve months' notice at any time. Aliant was right. Government regulators stated the last comma meant the last clause was independent and applied to both the original five-year term and the renewal terms.
>
> Source: Robertson G. Comma quirk irks Rogers. The Globe and Mail. 06 August 2006.

In both situations — courses and careers — your colleagues have the potential to improve the quality of your work. This chapter focuses on reviewing and evaluating documents and presentations.

6.1 Peer review

Peer review is the review of a colleague's work to provide feedback and improve the work prior to publication/presentation. Peer review occurs during the development phase of a document and/or presentation.

The purpose of peer review *is not* to make another person *feel better* by giving them false praise for their work.

The purpose of peer review *is* to help another person *be better* by giving them constructive recommendations to improve their work.

In the context of *Communicating Science*, peer review means endeavoring to help your colleagues be better communicators by providing them with specific suggestions to improve their work. If you receive a positive peer review with only a few minor recommendations, it should be because you have prepared a quality work.

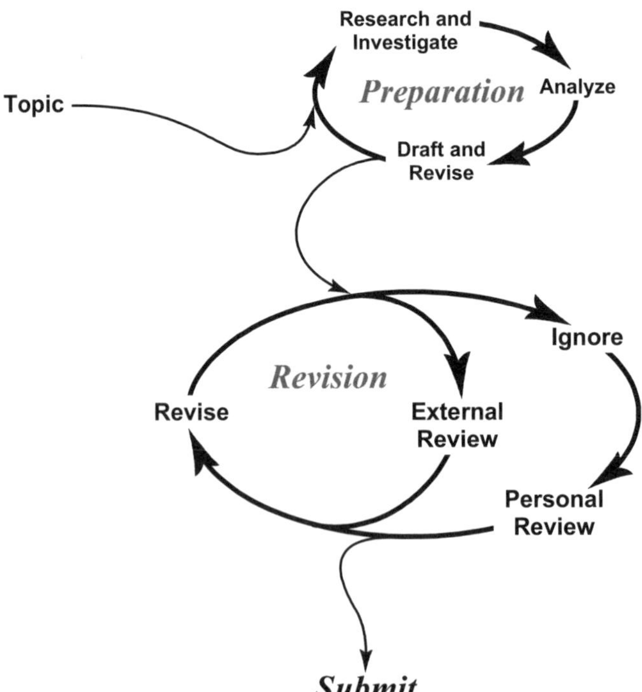

Figure 6.1 The preparation and revision stages involve numerous preparation cycles and numerous revision cycles before a document/presentation is complete.

During the preparation phase shown in Figure 6.1, you prepare a work that is complete in terms of content and you review the entire work a few times to ensure there are no obvious errors in content and flow. At this point, you are most willing to accept feedback from reviewers to improve the work. Consider the alternative: if you tweak your work until you perceive it to be "perfect" and then give it to others to review, you will be more protective of your work and less willing to accept feedback.

While others are reviewing your work, put the work aside. Review it again yourself when you receive the external reviews. You may be surprised that what made sense when you were working on the document does not make sense now.

Plan for a minimum of two review cycles during the revision phase, with one or two different reviewers each cycle. The reviewers should have a basic understanding of the subject so they can provide recommendations consistent with the discipline. Ideally, each cycle will uncover fewer errors and the feedback will increasingly fine-tune the work.

- *First review cycle*: print the document double-spaced. This gives the reviewers space in the document for extensive comments. For presentations, print two slides per page.
- *Final review cycle*: print the document as the properly formatted final document. The reviewers can then review the document text, layout, and formatting. For presentations, print two slides per page.

For posters and oral presentations, once your document has been reviewed a few times, have your colleagues review your presentation. Considerations for reviewing presentations are on page 222.

You are not obligated to incorporate the reviewers' recommendations into your work, but the reviewers did take the time to review your work and/or sit through your presentation and provide you with constructive feedback. A reasonable approach would be to accept their suggestions unless there is an obvious disconnect between their suggestion and the message you are trying to convey. Even if you choose not to accept their suggestions, they evidently misunderstood your work, so you must revise the work so that it conveys the intended message!

Peer review in a classroom

It is often uncomfortable to effectively engage in peer review. However, peer review is a valuable skill to learn because it produces better work, improves writing and presenting skills, and teaches interpersonal skills. While this may feel uncomfortable at first, a classroom is the *safest* place to learn and hone these skills.

Respect and professionalism are critical to establishing peer review as a valuable and positive experience. You, your fellow students, and the instructor must work to establish a positive environment conducive to peer review — one where everyone feels comfortable working with each other and speaking out in class. Below are some strategies for getting comfortable with peer review:

1. Start by reviewing third party documents in small groups and as a class. Role-play giving constructive feedback.
2. Review draft scholarly articles from research faculty, with their permission. The class is now engaged in a genuine and valuable task. Have the authors attend and discuss the recommendations.
3. Review presentations by the instructor or invited speaker.
4. Review documents and presentations by fellow students. Start with short documents, and move to larger documents and presentations.

Instructors: some course management systems have the ability to coordinate anonymous peer review similar to that done by scholarly journals. However, these systems do not provide the opportunity for in-person feedback, which is also a valuable skill.

Students: in addition to learning these skills, you are learning about additional topics within the discipline as you review each other's work. This is a good thing!

6.2 Editing

Editing is the process of reviewing a document to make the language

clear, coherent, concise, precise, and correct

Formally, there are several different types of editors.

- *Development editors* are involved in all aspects of document preparation, from making suggestions on what information to publish to large-scale writing and re-organization of the developing document. Development editors are common when completing large projects, such as books.
- *Substantive editors* organize and arrange the sections within the document to improve clarity; they may rewrite sections to improve readability and incorporate feedback from reviewers.
- *Copy editors* review the formatting, style, and grammatical structure of the document, but they do not check the information or substantively reformat the document.
- *Proofreaders* provide a final check of the document and identify spelling, grammar, style, and formatting errors not yet corrected.

In large organizations, you may encounter editors whose responsibilities cover one or more of the above roles. Professional editors and proofreaders are not specialists in the material they are editing — they are specialists in preparing quality documents. In large organizations, you and your colleagues may be called upon as reviewers. In smaller organizations, you and your colleagues will take on the roles of substantive editor, copy editor, proofreader, and reviewer.

In the publishing industry, editors will correct and format a document for publication. Your colleagues will not. Your colleagues will provide suggestions and it is your decision whether to incorporate their suggestions or not.

This section focuses on annotating printed documents because it is faster than annotating electronic documents and because it requires the author to consider each suggestion, determine if and how to incorporate the suggestion, and make the change to the master document.

> *Providing the author with suggestions, but allowing them to decide if and how to incorporate the suggestion, provides a mechanism for implementing peer review while addressing concerns of authorship and academic integrity. This strategy is consistent with the peer review process used by scholarly journals.*

Section B.5 provides strategies on annotating electronic documents.

Symbol	Meaning	Text with error	Corrected text
^ v	insert indicated punctuation or text.	In Febuary 2009 experiments testing the validity of Thomas model were	In February 2009, experiments testing the validity of Thomas' model were
#	insert space	sodium emission at 589.3nm	sodium emission at 589.3 nm
⊙	insert period	travelling at 100 km/hr	travelling at 100. km/hr
1/N	insert en dash	melting point of 155 157 °C	melting point of 155–157 °C
1/M	insert em dash	Sample 3 which should have	Sample 3 — which should have
ℓ	delete text	the laser was used to ionize the sample	the laser ionized the sample
⁀	close up spacing	an angle of 125 °	an angle of 125°
∽	transpose	live either in freshwater or in saltwater	live in either freshwater or saltwater
( ) ↗	move to indicated location	In case of fire, pull the fire alarm (immediately)	In case of fire, immediately pull the fire alarm.
¶	start a new paragraph	accurate under ambient conditions. The compressibility, Z, measures the	accurate under ambient conditions. The compressibility, Z, measures the
○	write out or abbreviate	use 2 250 mL beakers	use two 250 mL beakers
. . .	let it stand (ignore suggestions)	confirmed a negative reaction enthalpy.	confirmed a negative reaction enthalpy.

Symbol	Meaning	Text with error	Corrected text
/	make lowercase	0.10 mol/L Hydrochloric Acid	0.10 mol/L hydrochloric acid
≡	make uppercase	The large hadron collider in Europe	The Large Hadron Collider in Europe
=	make small caps	the molarity (M) of a solution	the molarity (M) of a solution
~	make **boldface**	let N be a 2 × 2 matrix	let **N** be a 2 × 2 matrix
—	make *italics*	E = m c²	$E = m\ c^2$
(rom)	make roman	The *largest change* was observed	The largest change was observed
(wf)	wrong font	The largest change was observed	The largest change was observed
(±pt)	change font size	The largest change was observed	The largest change was observed
⊏ ⊐	move left/right	$F = \ m\ a$	$F = m\ a$
⊓ ⊔	raise/lower	$C_1 V_1 = C_2 V_2$	$C_1 V_1 = C_2 V_2$
⊐ ⊏	center	The End	The End
‖	align	Hydrothermal vents on the ocean floor support an abundance of life.	Hydrothermal vents on the ocean floor support an abundance of life.
∧	superscript	3.00·108 m/s	$3.00 \cdot 10^8$ m/s
∨	subscript	0.25 mol/L CH3COOH	0.25 mol/L CH_3COOH

Editing notation

The previous page lists many of the common symbols used by editors and proofreaders to identify errors in a document. These symbols are made either in the text or in the margin, at the preference of the reviewer.

- *Editors* prefer in-text editing because the document has more errors at this stage of development.
- *Proofreaders* prefer in-margin editing because the document is formatted in final form and the document (hopefully) has fewer errors.

When editing another person's work, use a different colored pen — red or green — so your recommendations are visible.

In addition to editing and formatting text, a reviewer should also be reviewing the sentence and paragraph structure. Table 6.1 lists some common errors.

Table 6.1 Common word and sentence construction errors. See Chapter 1 for details. These symbols are circled, such as the font and font size recommendations on the previous page.

Symbol	Meaning	Symbol	Meaning
sp	spelling error	agr	agreement problem (subject/verb or pronoun/antecedent)
awk	awkward construction	mm	misplaced modifier
rep	unnecessary repetition	shift	change in tense/aspect/voice/ mood/person/number
rel	relevance?	||	incorrect parallelism
cs	comma splice	rom	roman font
frag	fragment	__ pt	__ point font
ros	run-on sentence	wf	wrong font

~~When teaching, we~~ use instructional strategies that mirror the learning ~~stage we prefer~~. (We teach the way we learn.) ~~If~~ an instructor and student have similar preferred learning ~~stages~~ the information being taught overlaps with the students existing knowledge~~, and~~ the student is interested in learning the material, ~~the student will find learning easier.~~ When there is a disconnect ~~between how the instructor teaches and how the learner learns or when the information is not linked to knowledge the learner already has,~~ learning will be more difficult. The ~~learner~~ can still learn, but will feel somewhat frustrated ~~in class~~. These are excellent opportunities to develop the other learning ~~stages.~~

Figure 6.2 An extract from an early review of *Communicating Science* during the preparation phase This is me — the author — reviewing my own work. The final text, which has undergone several additional revisions, is on page 122.

6.3 Giving and receiving feedback

One of the hardest things to do is *receive* feedback. Writers often take constructive suggestions as a personal attack and then defend themselves and their work. Even senior scientists, who have published dozens of research articles, may respond this way.

It takes self-confidence, a positive mindset, and practice to keep an open mind when receiving feedback. Learning how to give and receive feedback is critical to a positive peer review experience, quality work, and a long, productive career.

Realize that the person giving the feedback is probably not comfortable giving feedback either. *Work together and help each other!* The person receiving feedback can provide the reviewer with suggestions to improve how they give feedback.

Strategies for *giving* constructive feedback

Reviewers must provide constructive suggestions to improve the work. Saying "this is bad" alienates the reviewer from the author. When reviewing the document, identify

- strengths of the document
- areas that require revision

By identifying areas (sentences, paragraphs, figures, tables, sections) that are well written and explaining what is good about them, the writer is informed of strategies they should continue to use. In areas that require revision, the reviewer must explain why revision is recommended and provide one or more suggestions to improve the area.

When giving feedback, you do not need to go over every suggestion you provide. Focus your time on

- the strengths of the document (to maintain open communication)
- areas with several revisions (to explain your editing notes)
- the areas requiring significant revisions (to most improve the document)

Considerations when giving feedback:

1. Give feedback as soon as possible.
2. Hold the feedback discussion in private.
3. Start the discussion by asking the author to self-assess: what are their thoughts on the document? What sections is the author proud of? Where do they think more work may be necessary?
4. For each section, give positive observations first, followed by recommendations for improvement.
5. Don't be judgmental. Comments such as "This is poor" and "Have you even read this?" have no place in constructive feedback.
6. Use "I"-statements to take ownership of your feedback:
    - "I like …"
    - "I feel …"
    - "I noticed …"
    - "I found …"
    - "I believe …"
    - "In my opinion, …"
7. Take your time. Let the author think about your feedback before moving on. Ensure they understand what you are saying. Encourage them to ask questions at any time.
8. Don't argue if the author rejects a revision — they may be getting frustrated or you may be wrong. Move on to the next recommendation or take a break.
    - "Okay, mine is only a suggestion. We can move on or take a break if you like."
9. Don't overwhelm the author with too many suggestions. This is why there are multiple review cycles: each successive review session identifies increasingly minor errors.
10. Observe the emotions of the author. Are they frustrated? Angry? Emotional? Take a break if you perceive that they have stopped listening.
11. Don't share the review with others. Asking for feedback is akin to opening yourself to a personal attack. If the trust is broken, it is impossible to restore.

Finally, ask for feedback on your feedback! How could you have presented your suggestions better?

Strategies for keeping an open mind when *receiving* feedback

1. Constantly remind yourself that the feedback is for your benefit and the reviewer is trying to provide constructive feedback. Focus on positive emotions. Force yourself to smile and say "Thank you".
2. Meet with reviewers one at a time.
3. Pay attention. Do not interrupt the reviewer when they are talking.
4. Acknowledge when the reviewer says something that you agree with (both strengths and problem areas in your work).
5. Summarize the feedback you have received to ensure you understand it correctly. Ask questions if you do not understand the suggestions.
6. Don't take offense at the suggestions made by the reviewer. They are providing their opinion and constructive suggestions. If you disagree, ask for clarification, but do not openly argue with the suggestions.
7. Offer suggestions or advice to the reviewer, especially if they start making negative comments about your work.
8. Ask for a break if you feel yourself getting frustrated, angry, or emotional.
9. Don't discuss the feedback session with others.

Finally, ask for feedback on how the session went. Does the reviewer perceive you as attentive and accepting of feedback?

> *It is the mark of an educated mind to be able to entertain a thought without accepting it.* — Aristotle

6.4 Peer evaluation

Peer evaluation is the evaluation of a colleague's document or presentation against established criteria. Peer evaluation occurs after a document is submitted and/or during the presentation and is meant to critically and impartially assess a colleague's work.* This is an important skill you will need in your future studies and career.

Throughout your career, you may be required to train new employees and to evaluate your colleagues. You will also be expected to provide management with an impartial and justifiable assessment of their performance. You may also be meeting with sales representatives and assessing their credibility and their product. As you advance in your career, you will be promoted into supervisory and managerial roles. Some of your responsibilities may include interviewing prospective employees, conducting training sessions, conducting job evaluations, and evaluating reports and proposals. Being able to impartially evaluate your peers is a learned skill. Proficiency improves with practice.

Impartial evaluation requires standards. Appendix D contains evaluation rubrics for the different documents and presentations in *Communicating Science*, and contains an evaluation rubric to assess feedback.

Additional resources …

… on giving and receiving feedback

Receiving and Giving Feedback [internet]. University of Waterloo: [cited 12 May 2010]. Available from https://uwaterloo.ca/centre-for-teaching-excellence/teaching-resources/teaching-tips/assessing-student-work/grading-and-feedback/receiving-and-giving-effective-feedback

* Many people consider peer evaluation similar to the evaluation of scholarly documents.

Appendix A: Review questions

Use the following questions to guide the development and review of scholarly documents. Note that some questions apply to only some types of documents and presentations.

Content

- What do you believe the key points of the document are?

Layout

- Are there any spelling or grammar errors?
- Does the *Title* reflect the nature of the project?
- Does the *Abstract* summarize the key findings?
- Does the *Introduction* provide a comprehensive and balanced review of the existing knowledge in the field?
- Does the *Introduction* explain why this project was conducted?
- Is there sufficient information in the *Methods* section for another person to repeat the experiment(s)?
- Is there sufficient information in the *Results* section for another person to repeat the analysis?
- Are the *Discussion* and *Conclusion* supported by the data?
- Are there any technical terms that have not been explained?
- Are there any errors in the mathematical and/or chemical formulae?
- Are there any errors in the data, analysis, or interpretation?
- Do the tables and figures succinctly present the data? Are they referred to in the text? Are the captions self-explanatory?
- Are the references at an acceptable scholarly level and cited properly?

Meta-analysis

- Are the headings consistent throughout the document?
- Does the text follow a logical progression? Is there another way to organize the text to better convey the information?
- How could a reader possibly misinterpret the text?
- Is the text clear and concise, with unnecessary words eliminated?
- Is the work focused and on-topic? Is information presented in a positive and professional tone? Is it interesting and engaging?
- Is the societal importance of this project evident?
- Does the document achieve its purpose in a clear, coherent, concise, and precise manner?

Appendix B: Electronic document preparation

This appendix is an introduction to some of the functionality in word processing, spreadsheet, and presentation software, with a focus on the Microsoft Office® suite of products. This introduction assumes you have a basic understanding of Microsoft Word®, Excel®, and PowerPoint®. To use functionality not presented in this appendix, and to learn how to use other software, consult the Help and Tutorial features in that software or search the internet. It is likely that someone has needed the same functionality.

Table B.1 Keyboard shortcuts common to most productivity software.

<table>
<tr><th>Keyboard*</th><th>Text commands</th></tr>
<tr><td colspan="2">file commands</td></tr>
<tr><td>CTRL+N</td><td>new file</td></tr>
<tr><td>CTRL+O</td><td>open file</td></tr>
<tr><td>CTRL+S</td><td>save file</td></tr>
<tr><td>CTRL+P</td><td>print</td></tr>
<tr><td colspan="2">selecting and moving text</td></tr>
<tr><td><arrow keys></td><td>move through text, one character at a time</td></tr>
<tr><td>CTRL+<arrow keys></td><td>move through text, one word at a time</td></tr>
<tr><td>SHFT+<left|right arrow></td><td>select single character to the left or right</td></tr>
<tr><td>SHFT+<up|down arrow></td><td>select line of text above or below</td></tr>
<tr><td>CTRL+A</td><td>select all</td></tr>
<tr><td>CTRL+C</td><td>copy selected text and/or objects</td></tr>
<tr><td>CTRL+X</td><td>cut selected text and/or objects</td></tr>
<tr><td>CTRL+V</td><td>pastes the cut or copied text and/or objects</td></tr>
<tr><td>CTRL+F</td><td>Find dialog box</td></tr>
<tr><td>CTRL+H</td><td>Replace dialog box</td></tr>
<tr><td>CTRL+Y</td><td>repeat last operation</td></tr>
<tr><td>CTRL+Z</td><td>undo last operation</td></tr>
<tr><td>F7</td><td>check spelling</td></tr>
<tr><td>F1</td><td>help menu</td></tr>
<tr><td colspan="2">formatting selected text</td></tr>
<tr><td>CTRL+B</td><td>bold</td></tr>
<tr><td>CTRL+I</td><td>italic</td></tr>
<tr><td>CTRL+U</td><td>underline</td></tr>
</table>

* On Apple® computers, the CTRL key is the command key, ⌘.

B.1 Word processing (Microsoft Word®)

Table B.2 Keyboard shortcuts in Word and other common word processing software.

Keyboard*	Formatting Word® documents
CTRL+L	left-align paragraph†
CTRL+E	center-align paragraph†
CTRL+R	right-align paragraph†
CTRL+J	justify (left and right align) paragraph†
CTRL+SHIFT+S	change the style of the selected text or paragraph
CTRL+SHIFT+F	change the selected text's font†
CTRL+SHIFT+P	change the selected text's font size†
CTRL+SPACEBAR	remove paragraph or character formatting
CTRL+SHIFT+'+'	superscript
CTRL+'='	subscript

* On Apple® computers, the CTRL key is the command key, ⌘.

† You should not need these commands if you use styles to format your document.

View

Of the possible views of your document, two are most common:

- Print Layout: provides the layout of your document when printed
- Web Layout: provides the layout of your document as web content

For the documents described in Chapter 5, Print Layout should be used.

Word documents contain numerous formatting characters hidden in the text. The default mode is to show only the printed text. However, it is sometimes necessary to see these formatting characters, especially if there is a problem with the formatting. Press CTRL+SHIFT+8 to toggle between showing and hiding the formatting characters. Some common formatting marks include

space:	·	
tab:	→	
line break:	↵	(starts a new line within a paragraph)
paragraph:	¶	(end of paragraph)
special formatting:	▪	(seen for headings)
image anchor:	⚓	(only for floating images)
bookmark:	I *or* [ ]	
column break:	....................Column Break....................	
page break:	....................Page Break....................	
section break:	::::::::::::::::::::Section Break (<Type>)::::::::::::::::::::	

Additional information and strategies for a clean document:

- *Do not* use Enter to put space between paragraphs, use Spacing:After and Spacing:Before in the paragraph style (see below).
- *Do not* use multiple tabs to align information: → → → →<info> Set the desired tab stop in the style (see below).
- "Special formatting" is commonly found in headings and includes Keep with next, Keep lines together, and/or Page break before.
- Use Widow/Orphan control on all paragraphs.
- Hidden text is shown with a dotted underline.

Page setup

The Page Setup advanced settings (↘ icon) on the Page Layout ribbon allows you to set the paper size, paper orientation, and margins. You can control how these settings, headers, and footers apply to odd and even pages and the first page of each section.

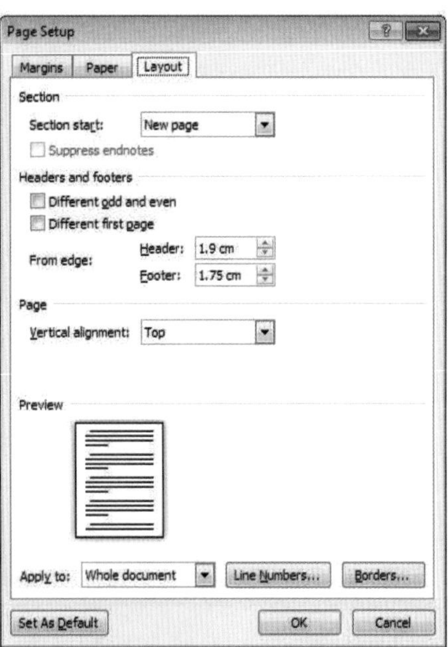

Figure B.1 Setting the page setup: margins and orientation (left), and section information and header/footer control (right). The Paper tab is not shown.

Sections

If you are making a complex document with multiple chapters, different types of page numbering, different headers or footers, different page orientations, or different numbers of columns, you will need to break your document into *sections*. The properties of each section can be adjusted independently. Select the Breaks drop-down box on the Page Layout:Page Setup ribbon to insert different types of breaks. To change the type of section break, use the Layout tab in the Page Setup dialog box.

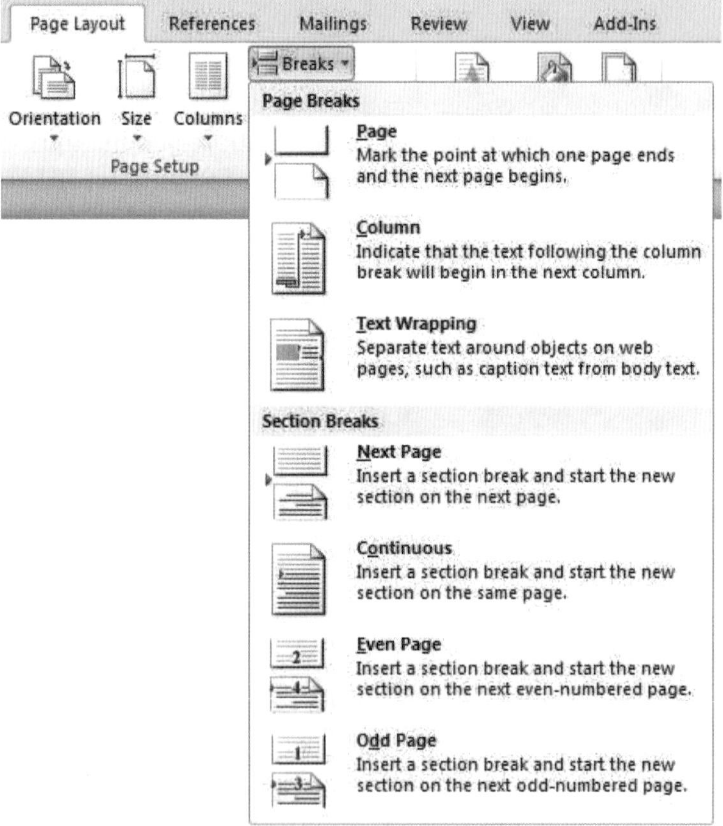

Figure B.2 The possible manual breaks that can be inserted into a document.

Styles

Styles control the formatting of paragraphs, tables, and text. Using styles enables you to control the layout of all the paragraph/tables/text from one location. For example, changing from Arial to Times New Roman font, or adjusting the spacing around paragraphs and headings are all controllable using styles. Common styles are shown on the Home ribbon.

If an organization or journal provides a template with the styles pre-defined, use it! Otherwise, use this section to create your own template.

The Styles advanced settings allow you to view, edit, and create styles. Hover over the style to see the configuration. Right-click and select Modify to configure the style to your requirements.

- *Paragraph* styles format both the paragraph and all the text within it. Common paragraph styles include Normal, Heading 1/2/3, List bullet 1/2/3, List number, Table/Figure captions, Header, and Footer.
- *Table* styles format the information in tables.
- *Character* styles format selected text within a document. Select text using the SHFT+arrow keys or use the mouse.

For ease in reading and formatting, it is best to use a limited number of styles. If you use styles properly, you should not have to use the Font and Paragraph dialog boxes on the Home ribbon. Proper use of styles also allows you to dynamically create a table of contents, table of figures, etc.

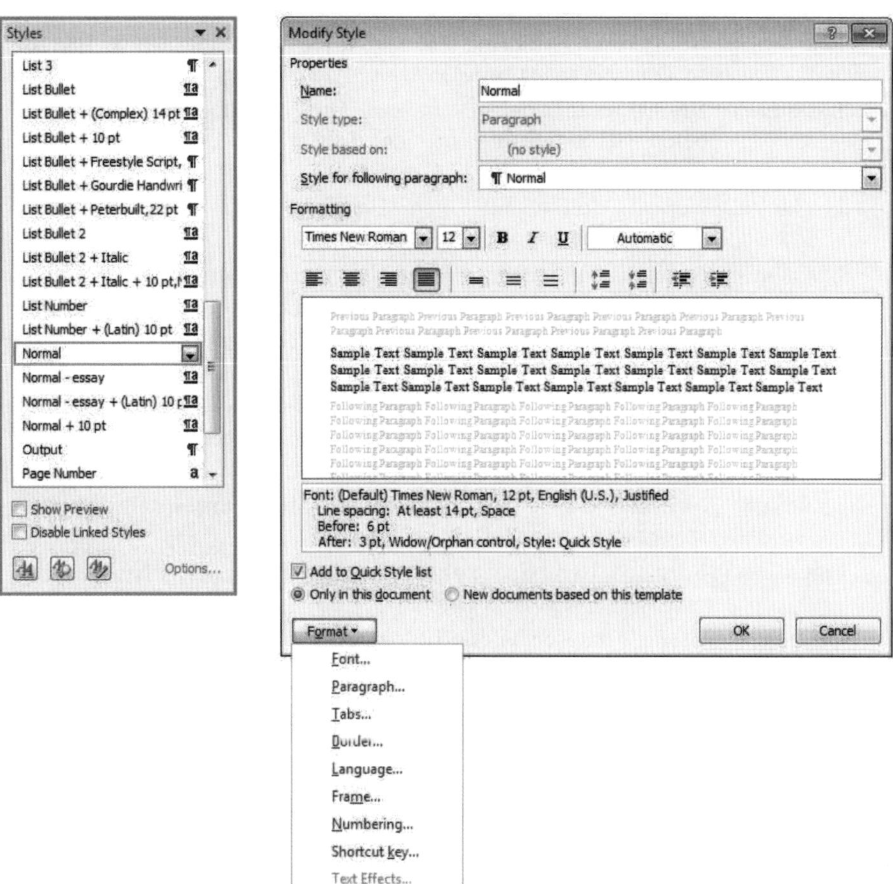

Figure B.3 The Styles advanced settings and modification dialog boxes.

Formatting of the main styles in *Communicating Science*

Paragraph styles

Heading 1: Normal + 16 pt bold. Space before: 0; space after: 12. Page break before. Single line spacing. Keep with next. Level 1 outline numbering. Numbered as 'Chapter #.'

Heading 2: Normal + 14 pt bold. Hanging: 1.0 cm. Space before: 18 pt; space after: 9. Single line spacing. Level 2 outline numbering. Numbered as '#.#' (Level 1.Level 2)

Heading 3: Normal + 12 pt bold. Space before: 12; space after: 3. Single line spacing.

Heading 4: Normal + 12 pt italic. Space before: 9; space after: 3. Single line spacing.

Normal: 12 pt Times New Roman font. Space before: 3; space after: 3. Justified. Line spacing: at least 14 pt. Widow/Orphan control.

List: Normal + left indent: 0.5 cm. Space before: 0. Single line spacing. No space between same style paragraphs.

List bullet: Normal + left indent: 0.2 cm. Hanging: 0.3 cm. Space before: 0. Single line spacing. Tab at 6.25 cm, centered. Bulleted: level 1.

List bullet 2: Normal + left indent: 1.0 cm. Hanging: 0.3 cm. Space before: 0. Single line spacing. Bulleted: level 2.

List number: Normal + left indent: 0.3 cm. Hanging: 0.6 cm. Space before: 0. Single line spacing. Numbered: level 1.

Header: Normal + 10 pt. Space before: 0; space after: 0. Single line spacing. Tab at 6.25 cm, centered.

Example: Normal + 10 pt + left indent: 0.5 cm. Space before: 0; space after: 6. Single line spacing. No space between same style paragraphs.

Table cap.: Normal + 10 pt. Space before: 12. Single line spacing. No space between same style paragraphs.

Figure cap.: Normal + 10 pt. Space after: 12. Single line spacing. No space between same style paragraphs.

Quotation: Example + space before: 12; space after: 18.

Character styles

Emphasis: Font: italic.

Definition: Font: bold, italic.

Command: Font: 11 pt Arial.

Additional characters and symbols

Tables 1.1 and 1.2 list additional characters commonly used in scientific communication. To insert these characters into your document, select the Symbols:More Symbols... icon on the Insert ribbon. This brings up the character map shown in B.4. In addition to inserting characters from this dialog, you can alternatively use the shortcut key indicated for the character.

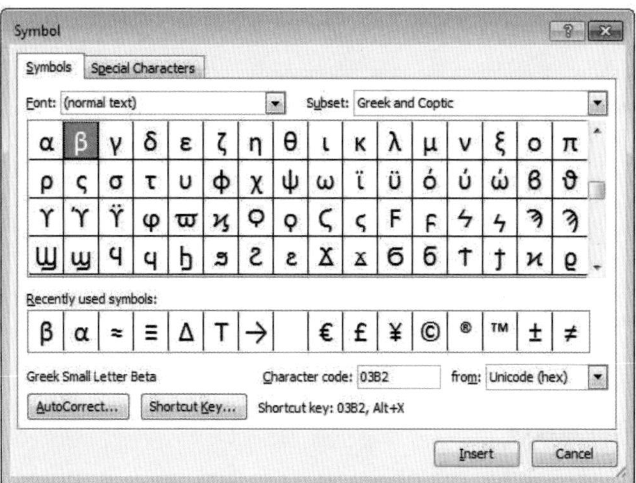

Figure B.4 The dialog box for inserting additional characters and special characters in Word. The default shortcut key is at the bottom center: 0382, Alt+X

Fields

Fields are a powerful but complex way to make a professional-looking document. The fields insert information about the document (such as the page number, author, filename, creation date, and current date) and information from other parts of the document (such as heading numbers and heading text).

The Quick Parts:Field... dialog box on the Insert ribbon allows you to directly insert fields into your document. In many cases, other dialog boxes also insert fields. For example, page numbers, bookmarks, tables of contents, and indexes are fields inserted from other dialog boxes.

Many fields have options that must be set to display the proper information. Learning these will aid you in preparing professional documents. Consult the Help menu in Word for assistance in inserting and formatting the information you desire. When you have a field selected, pressing ALT+F9 toggles between the field output and field code, allowing you to edit them.

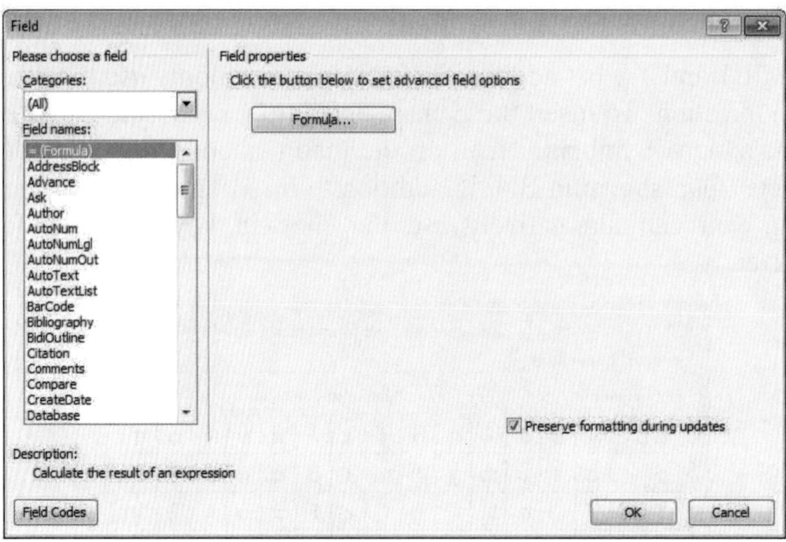

Figure B.5 The Field dialog box. The list of possible fields is in the scroll region on the left.

Headers and footers

Every page has space at the top (header) and bottom (footer) for information that is constant throughout the section or entire document: page number, document name, your name, copyright information, section name, ….

Figure B.1 (page 263) shows that it is possible to have different headers and footers on odd and even pages and to have a different first page header and footer. By default, headers and footers keep the same properties as the previous section. This is usually a good thing, but must be turned off to change the headers and footers, such as changing the page number format: {i, ii, iii, iv, …} → {1, 2, 3, 4, …}

Selecting Header:Edit Header in the Insert ribbon opens the Header & Footer Tools ribbon.

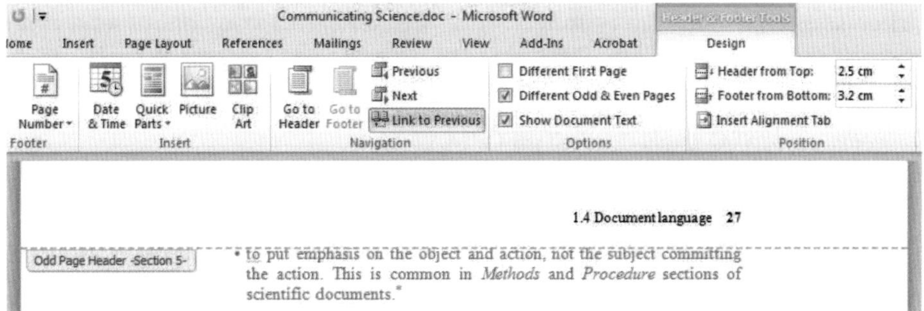

Figure B.6 An odd-page header from Section 7 of *Communicating Science*. Hovering over the toolbar illustrates some of the fields that can be inserted: page number, date, etc. The colored Link to Previous icon indicates that the Section 7 heading is the same as the previous section heading.

To incorporate section headings into headers (or anywhere else in the text), use the StyleRef field.

The even page heading of *Communicating Science* inserts the page number and Heading 1 with the field codes

output: **2** Communicating Science
code: { **PAGE** } { STYLEREF "Heading 1" \t * MERGEFORMAT }

The odd page heading of *Communicating Science* inserts Heading 2 and the page number with the field codes

output: 1.1 Elements of effective communication **3**
code: { STYLEREF "Heading 2" \n \t * MERGEFORMAT }.
{ STYLEREF "Heading 2" \t * MERGEFORMAT } { **PAGE** }

The \n field code inserts the label from the selected style: “Chapter 1”. The \t field code suppresses all non-numeric text: the word “Chapter”. Pressing SHIFT+F9 toggles between the field output and field codes.

Autonumbering

Tables and figures are numbered in academic works. The Seq field increments a number every time the field is inserted into a document. Most works have separate sequences for tables, figures, and equations.

Table 1, Table 2, …
Figure 1, Figure 2, …

To start, select Quick Parts:Field… on the Insert ribbon and then scroll to Seq. A simple autonumbering sequence called “Table” looks like

SEQ Table

After selecting OK, the number "1" is inserted into your document. You can toggle between the field output and field code by pressing SHIFT+F9. The actual field code is

code: { SEQ Table * MERGEFORMAT }

Copy and paste the caption, then modify the caption text, for all subsequent tables. To update the numbering, select the entire document, CTRL+A, and then press F9.

To reference the table sequence in your document, select Cross-reference on the Insert ribbon, then select the Table reference type and the reference you wish to cross-reference. Be careful that you insert the information you desire: entire caption, label and number, page number, etc.

Table 1 contains the titration data.

If you are preparing a document with multiple chapters and appendices, you may wish to prepend the table number with the chapter/appendix number and you may wish to restart the sequence every chapter. All the figures and tables in *Communicating Science* have this format. To insert the current chapter number, the StyleRef field is used and field codes used to control what is inserted.*

output: Table 1.1: <caption for Table 1>
code: Table { STYLEREF "Heading 1" \n \t * MERGEFORMAT }.{ SEQ Table \s 1 * MERGEFORMAT }: <caption for Table 1>

Note that the table field codes have an additional \s #.

code: { SEQ Table \s 1 * MERGEFORMAT }

The \s # field code restarts the Table sequence after passing Heading #. In *Communicating Science*, Heading 1 is used for chapters.

Table of contents/figures/tables

It is possible to have Word dynamically create tables of information — Table of contents, Table of figures, Indexes, etc. — from headings and field codes inserted into the document. These tables may be created using Table of Contents and Table of Figures icons on the References ribbon. The number of headings to display and formatting of the table of contents can be set from the Table of Contents page. The Table of figures and Table of tables can be created using the Table of Figures tab. The Options... button allows you select the style from which to build the table.

* Consult the Word Help menu for information on these and other field codes.

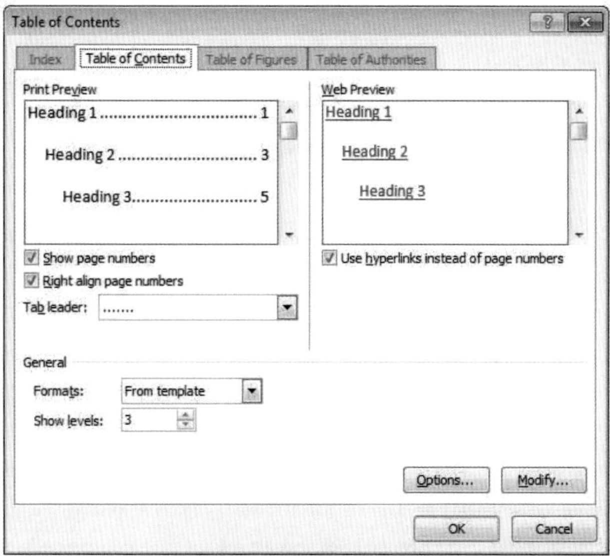

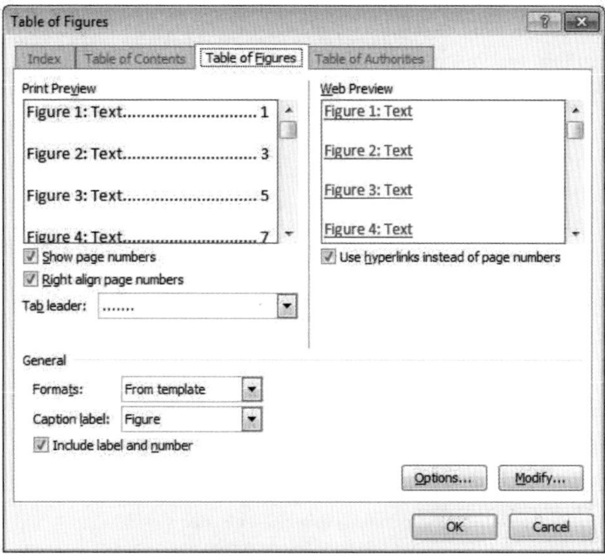

Figure B.7 Inserting a Table of contents (top), and Table of figures/tables (bottom).

Footnotes

The Insert Footnote and Insert Endnote icons on the References ribbon allow you to insert a footnote (bottom of page) or endnote (end of section or document) at the cursor location. By default, Word inserts separators between the text and the footnote. To change or remove these separators, select Draft on the View ribbon and then select Show Notes on the References ribbon. Figure B.8 shows what you can change.

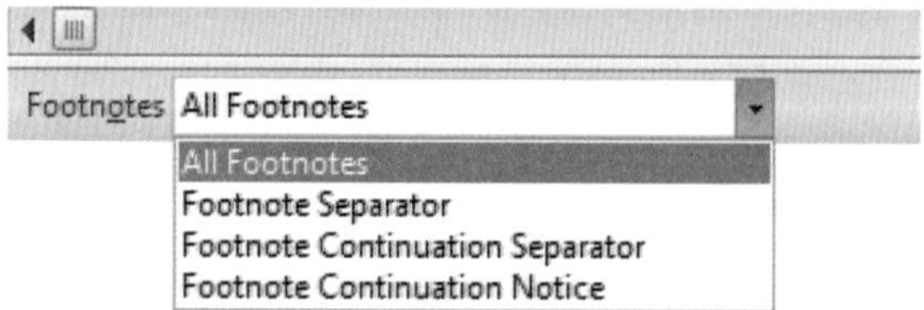

Figure B.8 The dialog box to change how the footnotes are separated from the text.

Spelling and grammar analysis

Before conducting a spelling and grammar analysis of your document, you need to configure the analysis at File:Options:Proofing. Adjust the settings to your preference or to the requirements of the instructor, employer, or publisher to whom you are submitting your work.

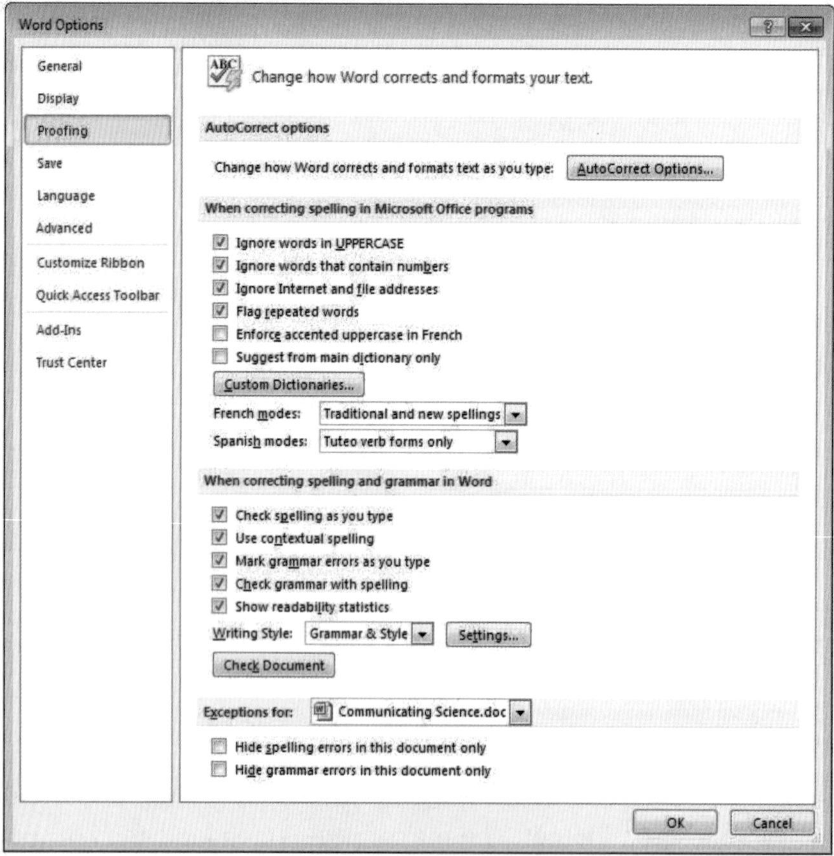

Figure B.9 The Proofing dialog box for configuring the spelling and grammar settings, and grammar rules.

Ensure that the Show readability statistics button is checked. After a grammar check, a dialog box shows a readability analysis of your document. Table 5.1 provides information on interpreting the readability statistics.

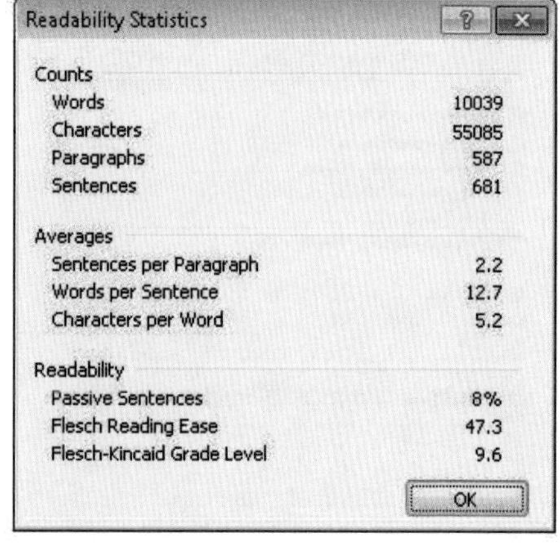

Figure B.10 Readability statistics of *Communicating Science.*

Note that the spelling analysis will not catch wrong words that are also words, which is why reviews are important.

breach, breech
if, in, is, it
meat, meet
spell, spill
than, that, then
wear, were, where
weather, whether
yore, your, you're

Dew knot putt awl yore trussed inn spill chequers.

Additionally, do not automatically accept all of the corrections suggested by spelling and grammar analysis. The software is often unaware of scientific terms and phrases.

Watermark (background)

You can insert a color, picture, or text background using the Watermark: Custom Watermark… advanced settings on the Page Layout ribbon. Text backgrounds could be a word like DRAFT, CONFIDENTIAL, COPY, or whatever text you want.

Autocorrect

The default in Word is to automatically apply formatting as you type, which becomes frustrating as you attempt to maintain consistency and a limited number of styles in your document. You should turn off most of the automatic formatting in order to retain control of the document. These settings are in AutoCorrect Options… dialog box on the File:Options:Proofing dialog.

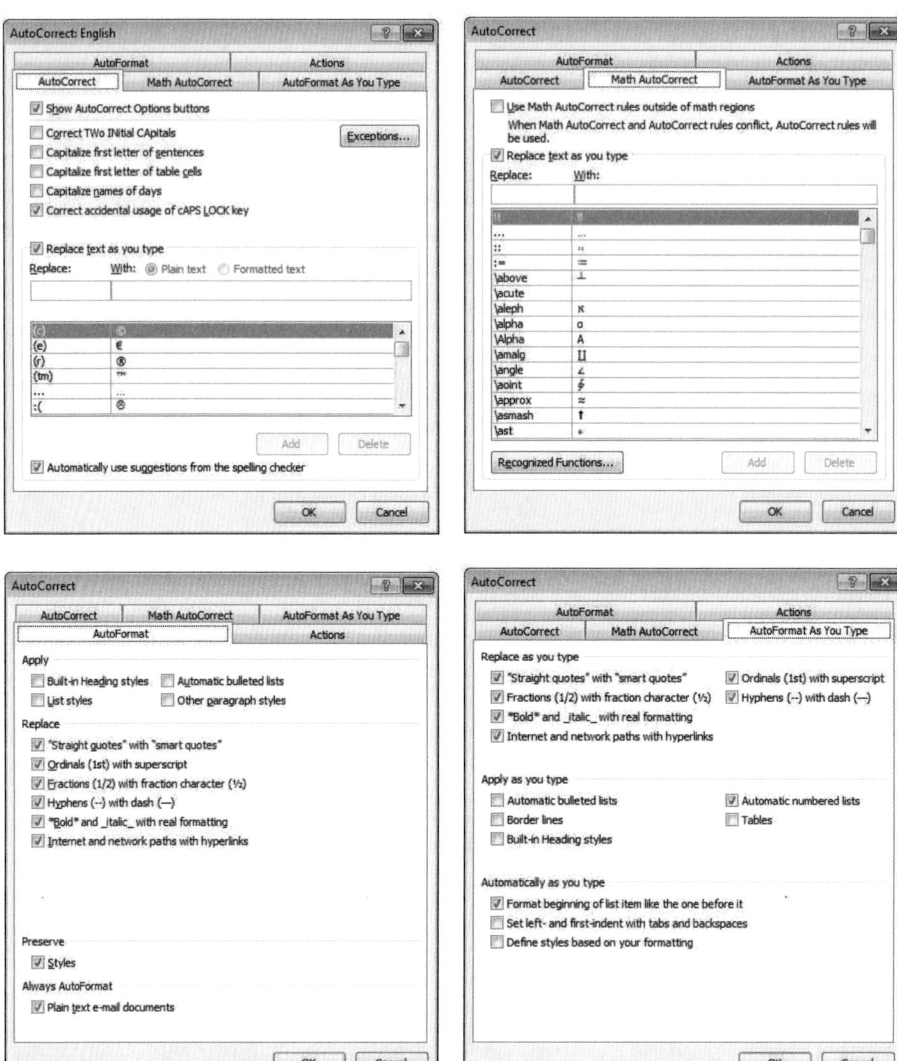

Figure B.11 Recommended autocorrect and autoformatting settings to give you the most control over the formatting of your document.

Autocorrect is valuable in that you can create custom corrections to automatically replace unformatted text with formatted text as you type. This is an excellent time-saver when having to repeatedly type a complex phrase or a phrase with specific formatting. When creating the correction, ensure

- the text you replace is not a common word or phrase
- you select With Formatted Text when creating the correction

Some sample custom corrections are given below.

dH → $\Delta_r H°$
GT → Green's theorem
fff → drosophilae

When you type the custom phrase and then press the Spacebar, Word automatically inserts the formatted text.

Templates

Setting all of the above parameters is a challenging and tedious task. Once you have everything configured properly, save the document as a template into the template directory. Once done, when you select File:New, your document will be one of the listed templates. If you want your document to be the default, save it as normal.dotx.

Preparing a book

Preparing a book-style work — an essay, thesis, technical report, manual, book, etc. — is an involved and time-consuming process. It is also a valuable learning experience and provides you with technical skills that you can use to improve all your documents. *Communicating Science* was created in Word. The following suggestions will help you create a professional-looking document.

- Use Next page, Odd page, and Even page section breaks to separate the front material from the body, to get the correct pagination between chapters, and anywhere else section breaks are needed to get the correct formatting. For example, Continuous section breaks allow you to change the number of columns of text in a section of the document.
- Modify the headers so that each section has the appropriate headings.
- Use styles to control the formatting of all your text.
- Use styles to dynamically create and update your table of contents, table of figures, index, etc.
- Use autonumbering to number chapters, figures, tables, equations, etc.

You may need to consult the Word Help menu or additional resources to learn how to configure Word as you prepare a book. If there is something you want to do in Word, it is likely possible and an internet search will tell you how to do it.

B.2 Spreadsheets (Microsoft Excel®)

Table B.3 Keyboard shortcuts in Excel and other common spreadsheet software.

Keyboard*	Formatting Excel® documents
F2 (Mac: ⌘+U)	enter cell to edit text
TAB	move to next cell in the row
SHFT+TAB	move to previous cell in the row
ENTER	move to the next cell in the column
SHIFT+ENTER	move to the previous cell in the column
CTRL+<left|right arrow>	go to end of text cells/next cell
SHFT+<arrow keys>	select cells

* On Apple® computers, the CTRL key is the command key, ⌘.

Page setup

The Page Setup dialog is similar to Word's, but there are additional tabs to set the header and footer and to define what sections of the sheet to print and if any rows/columns are repeated on every page. This is convenient if you have data that spans multiple pages.

Cells and cell formatting

Each worksheet contains a series of ***cells***. Each cell is uniquely identified by a column letter and row number. The top left cell is A1.

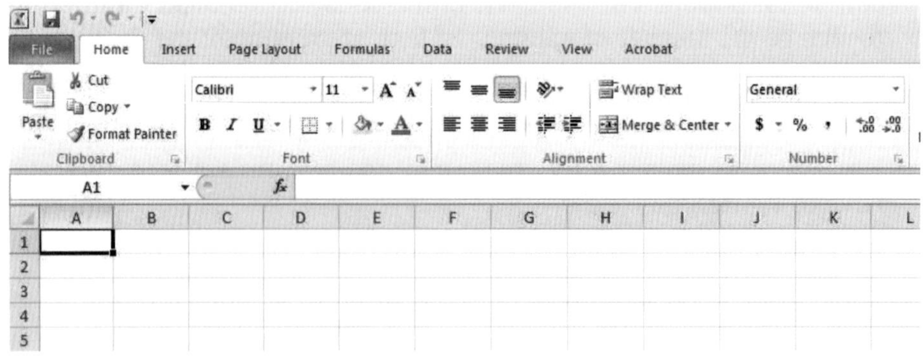

Figure B.12 A blank Excel spreadsheet.

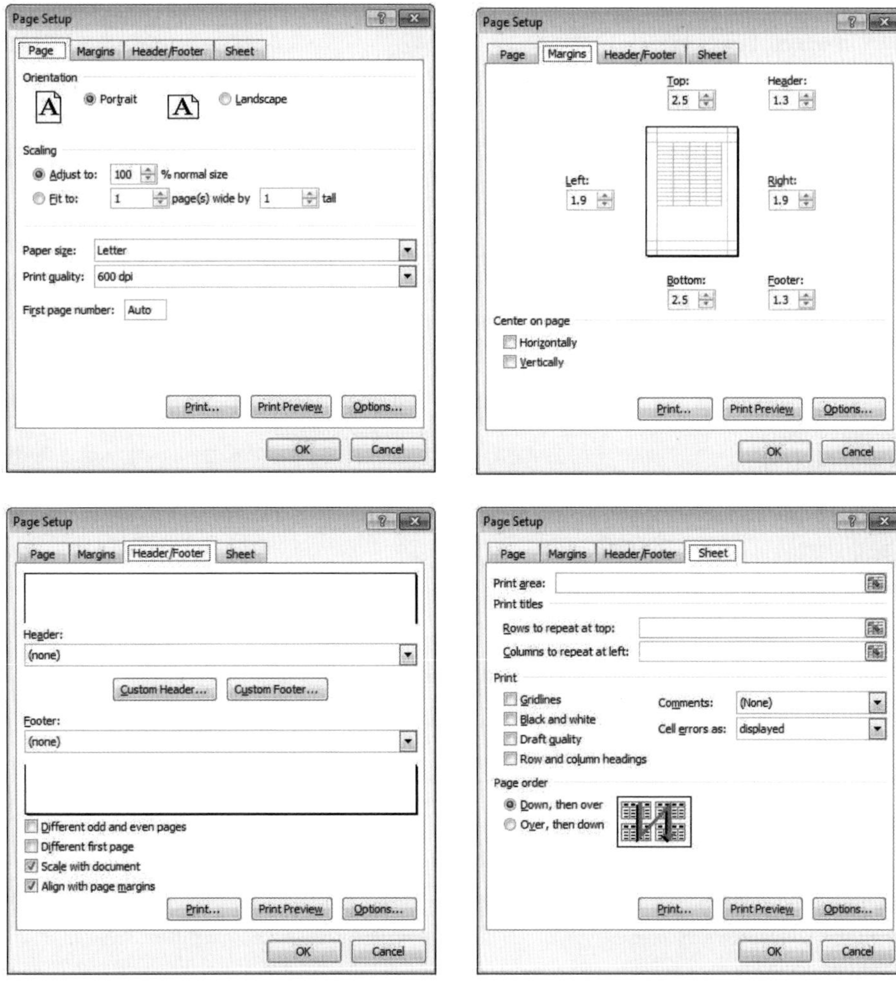

Figure B.13 The tabs in the Page Setup dialog box.

To format cells, select the cell(s) you wish to format and press CTRL+1 (or select the Font advanced settings on the Home ribbon). As illustrated in Figure B.14, it is possible to format

- how numbers are displayed, from the number of decimal points to scientific notation to currency
- the alignment of text and merging of cells
- thc font
- borders and shading of the cells

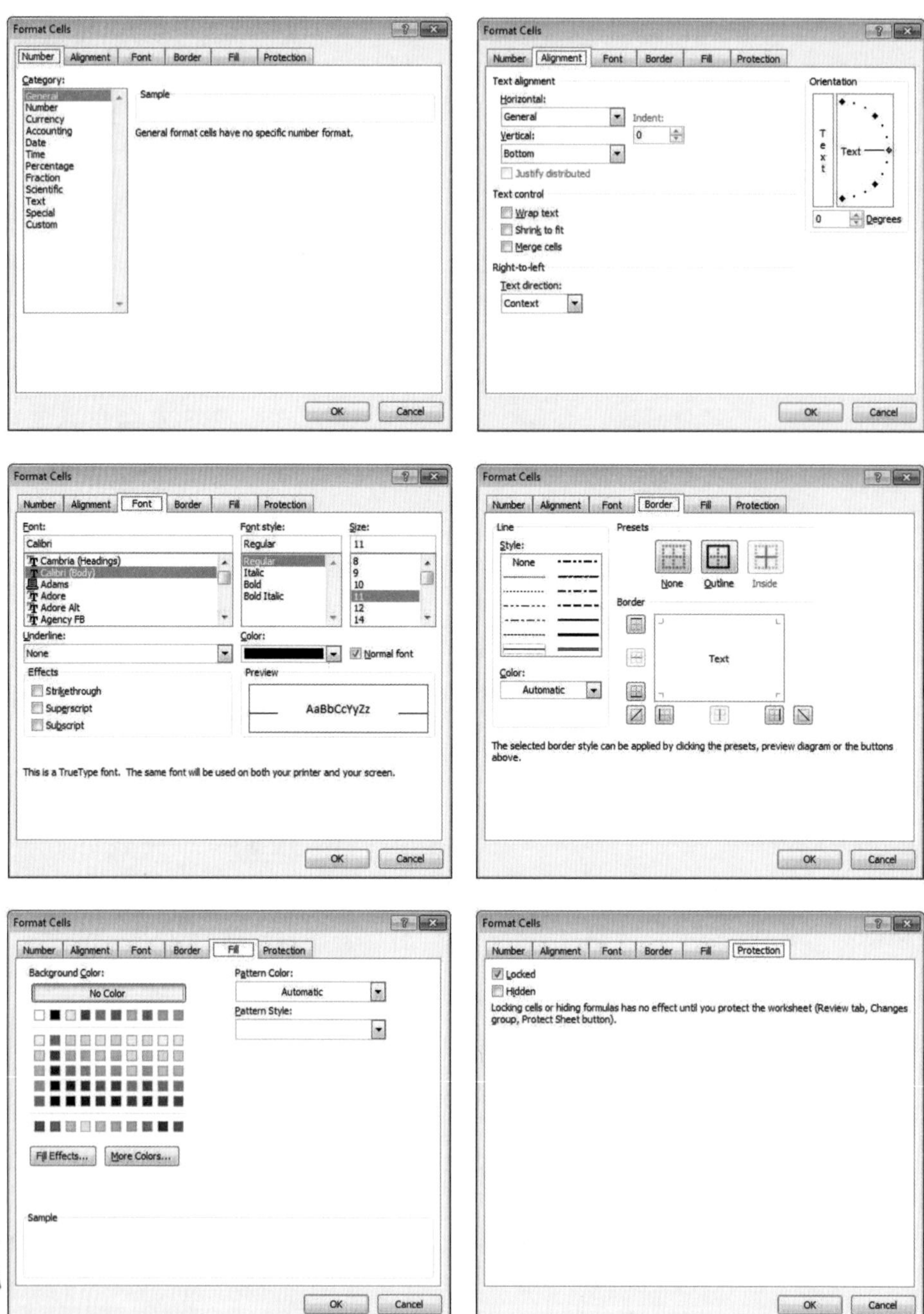

Figure B.14 The tabs in the Format Cells dialog box.

To format selected text within a cell (such as italicizing, subscripting, and superscripting), enter the cell, select the text, and then press CTRL+1.

Rows and columns can be resized and hidden. Figure B.15 shows how to format selected cells. This method of setting the row height and column width ensures the selected rows and columns are all the same size. It is convenient to hide rows and/or columns containing intermediate calculations when producing publication-quality tables.

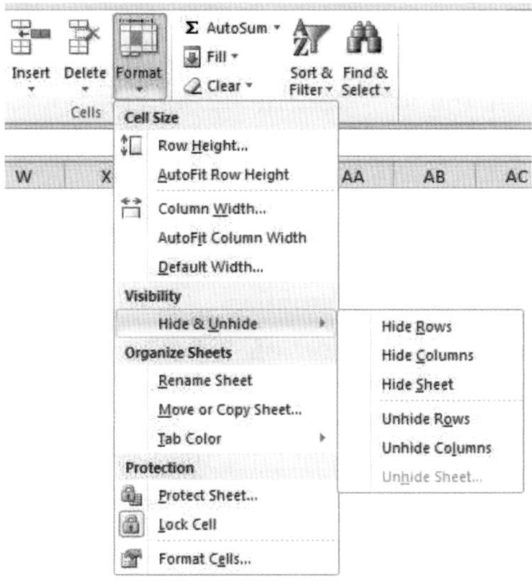

Figure B.15 Formatting rows and columns.

Publication-quality tables and figures can be created in Excel. All the tables and most figures in *Communicating Science* were created in Excel.

Inserting functions

Excel is designed to do math, from simple mathematical operations to advanced statistical analysis. Excel has hundreds of built-in formulae. Addition, subtraction, multiplication, and division are entered manually. Advanced formulae can be entered manually or using the Insert:Function icon on the Formulas ribbon.

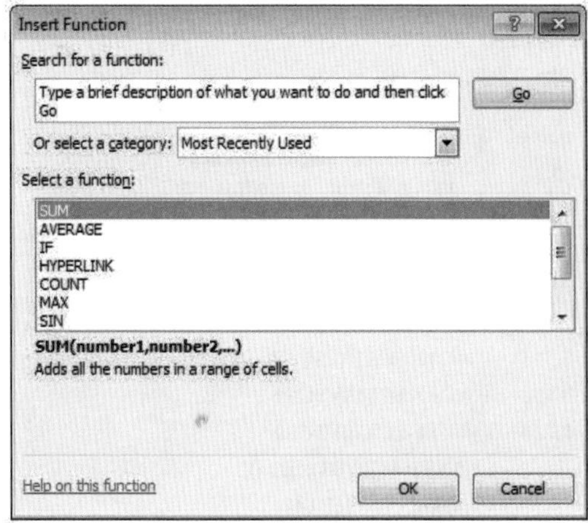

Figure B.16 Insert function dialog box.

In the cell, the formula starts with an equals sign, "=".

=A1*B1 =A1/B1

=SUM(A1:A4) =AVERAGE(A1:A4)

Once a formula has been entered once, it possible to replicate the formula for the other data in the series. The default action is that the cell reference moves *relative* to the cell. To make a formula reference a specific cell, such as a constant, no matter where the formula is copied,

- use a dollar sign in front of the letter to fix the column: $A1
- use a dollar sign in front of the number to fix the row: A$1
- use a dollar sign in front of both to fix the absolute position: $A$1

There are several categories of formula:

- *mathematical functions*: mathematics, financial, statistical calculations
- *array functions*: matrix calculations
- *logical functions*: results depend on the content of other cells
- *lookup functions*: matching and finding data
- *text functions*: parsing and formatting text

File Home Insert Page Layout Formulas Data Review View Acrobat

Cut Copy Format Painter Paste Clipboard | Times New Roman 10 Font | Wrap Text Merge & Center Alignment

M31 fx

	A	B	C	D
1		**Fe^{3+} data**		
2				
3	**label**	**Fe^{3+} /(mol/L)**	$A_{450\ nm}$	
4		*standards*		
5	std. 1	$6.73 \cdot 10^{-5}$	0.593	The standards data and the unknown absorbance are entered. The **slope**, **intercept**, and **unknown concentration** are calculated by Excel.
6	std. 2	$5.07 \cdot 10^{-5}$	0.475	
7	std. 3	$3.39 \cdot 10^{-5}$	0.336	
8	std. 4	$1.68 \cdot 10^{-5}$	0.199	
9		slope:	**7845**	**=SLOPE(C5:C8,B5:B8)**
10		intercept:	**0.070**	**=INTERCEPT(C5:C8,B5:B8)**
11		*unknown*		replicate absorbance measurements
12		**$3.25\underline{0} \cdot 10^{-5}$**	0.325	**=(C12-$C$10)/$C$9**
13	unk. A	**$3.35\underline{2} \cdot 10^{-5}$**	0.333	**=(C13-$C$10)/$C$9**
14		**$3.28\underline{8} \cdot 10^{-5}$**	0.328	**=(C14-$C$10)/$C$9**
15	average:	**$3.30 \cdot 10^{-5}$**		**=AVERAGE(B12:B14)**
16	std. dev.:	**$5.1 \cdot 10^{-7}$**		**=STDEV(B12:B14)**
17				

Figure B.17 Screenshot showing the formulae for several calculations: slope, intercept, average, and the unknown concentration.

Solver

Solver is a convenient add-in that fits your data to any equation you can enter into Excel. Solver adjusts the variables in your equation to minimize the difference between the data and equation.

Figure B.18 shows how the non-linear data in Figure 2.14 is fit to an exponential function, with the rate constant, *k*, being the variable that is adjusted by Excel.

File | Home | Insert | Page Layout | Formulas | Data | Review | View | Acrobat

Q30 | f_x

CO_2 dissolution data

	A	B	C	D	E	F	G
1							
2							
3	**Time /min**	***P*$_{CO2}$ /bar**		***P*$_{CO2}$(fit)**	**diff.**		
4	0.0	1.01		1.01	0.000	→ =B4*EXP(-$E$17*A4)	
5	0.5	0.96		0.98	0.014	→ =B5*EXP(-$E$17*A5)	
6	1.0	0.95		0.94	-0.011	→ =B6*EXP(-$E$17*A6)	
7	1.5	0.90		0.91	0.008	→ =...	
8	2.0	0.88		0.87	-0.006		
9	3.0	0.83		0.81	-0.017		→ =...
10	4.0	0.75		0.76	0.012		→ =D10-B10
11	5.0	0.70		0.70	-0.001		→ =D11-B11
12	7.5	0.59		0.59	0.001		→ =D12-B12
13	10.0	0.48		0.49	0.006		→ =...
14	15.0	0.34		0.34	0.004		
15	20.0	0.25		0.24	-0.009		
16	experimental data			SUMSQ =	*0.0010*	→ =SUMSQ(E4:E15)	
17				*k* =	*0.0726*	→ varied by Excel	
18							

Figure B.18 Screenshot showing the formulae for fitting an exponential function to the experimental pressure.

In Figure B.18, columns A and B are the experimental data. Column D is the pressure calculated assuming an exponential decay, given by

$$P_{CO_2} = P^0_{CO_2}\ e^{-kt}$$

Where *k* is the rate constant for this reaction and located in cell E17.

Column E shows the difference between the calculated and experimental pressures. Cell E16 is the sum-squared differences: the values in E4 to E15 squared, then added together.

Selecting Solver on the Data ribbon brings up the Solver dialog. (If Solver is not present, select File:Options:Add-Ins to activate the add-in.) In Figure B.19, Solver is set to vary cell E17 (the rate constant) to get a minimum in E16. This results in a non-linear function that best fits the experimental data. The guess value in E17 must be reasonable, and additional constraints can be added if required. Note that this example was simple, with only one varied value; it is possible to have Solver simultaneously vary several parameters in more complex functions to obtain the best non-linear fit to experimental data.

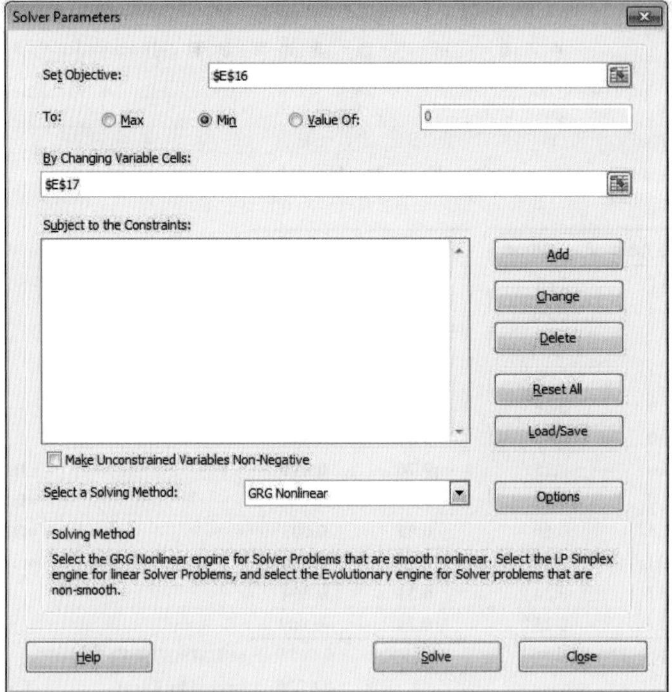

Figure B.19 The Solver dialog box.

Creating charts

Most scientific graphs plot *y* versus *x*. Excel calls these Scatter charts.

1. Select the data and then open the Chart dialog box on the Insert ribbon.
2. Select X Y (Scatter), or the chart type you require. Also select how you want the data displayed: points, straight lines, curved lines, or a combination thereof.
3. Select OK to create the chart.

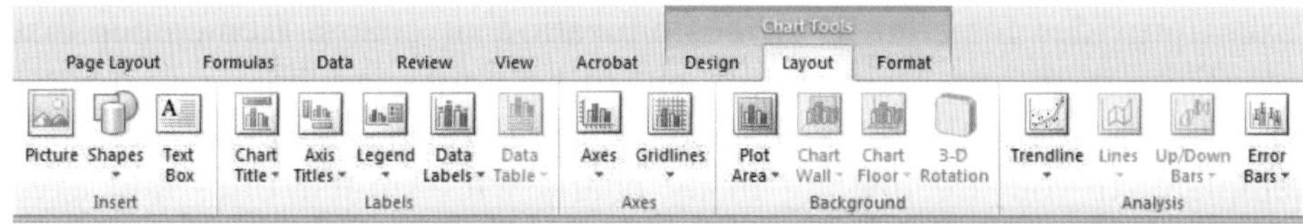

Figure B.20 The Chart Tools:Layout ribbon showing the icons for formatting a chart once created.

Make the following modifications to transform the chart into a scientific graph.

- Right-click the chart and Move Chart... to a new sheet. This provides you with more control over the size of the chart.
- Set the page margins to get the desired chart size. Start with 3.0 cm all around and 2.0 cm header and footer margins.
- Use the Labels and Axes icons on the Chart Tools:Layout ribbon to add axes (required) and a title (if required).
- To add additional data sets, right-click the chart and Select Data....
- Double-click on a data point to format the data series.
- Right-click on the title, axes values, axes labels, or legend to format the font type and size.
- Double-click on the axes, gridlines, or plot area to format the line size and style.

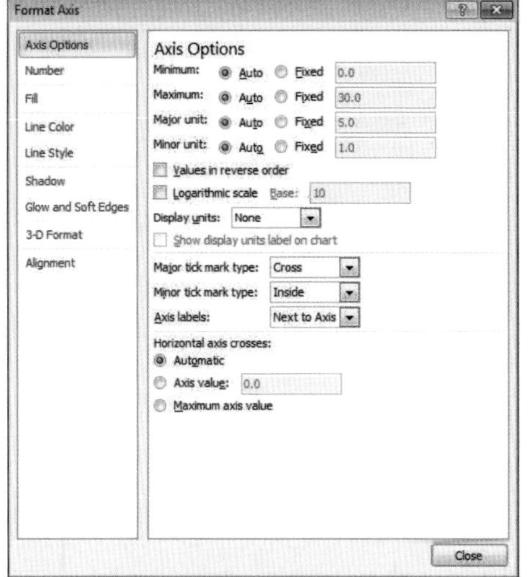

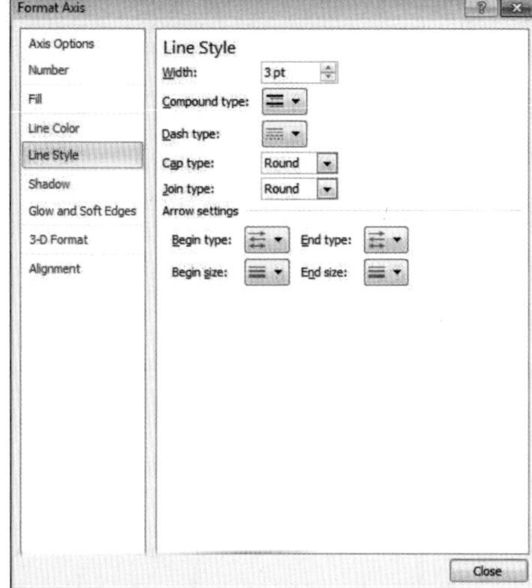

Figure B.21 Selected tabs of the Format Axis dialog box.

Excel provides for significant customization of charts. Remember that simplicity is an important component of scientific graphs. Additionally, once you have configured a chart to your preferences, save the chart as a template. If you are making many charts with similar titles and axes labels, you can duplicate an existing chart, change the data, and modify the axes labels.

The graphs in *Communicating Science* were created using the parameters below. The values you use will vary, depending on your document.

- text formatting
    - *title*: 24 pt bold Times New Roman
    - *axis label*: 20 pt bold Times New Roman
    - *axis numbers*: 20 pt Times New Roman; minimum number of significant digits (not the number of digits in the data; see the footnote on page 96)
    - *text on chart*: 16 pt Arial
- data points and lines
    - *border lines of chart and axis tick marks*: 3.0 pt (triple width)
    - *data points*: 10 pt
    - *line size with data points*: 1.0 pt (single width)
    - *line size without data points*: 2.0 to 3.0 pt (double or triple width)

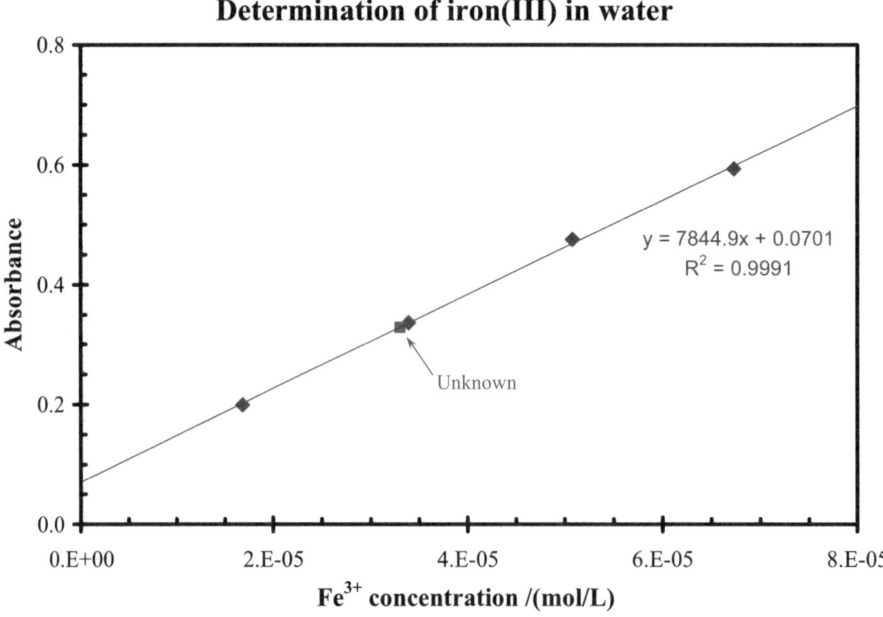

Figure B.22 Finished graph for inclusion in *Communicating Science*. Publication-quality graphs commonly have the title suppressed. (See Figures 2.9 and 2.10.)

Pasting from Excel into other documents

When pasting a table or chart from Excel, the default is to paste as an HTML table that is editable in Word and PowerPoint. This does not keep the formatting of your Excel table and is not the best option for pasting tables and charts! To keep the formatting, the best option is to choose Paste Special and paste as a Picture (Enhanced Metafile). This pasting, however, does have a few nuances:

- You must hide gridlines before copying.
- You may need to copy an adjoining column or row to get the proper line widths. (This bug is new to Excel 2007 and 2010.)

B.3 Presentations (Microsoft PowerPoint®)

Table B.4 Keyboard shortcuts in PowerPoint and other presentation software.

<table>
<tr><th>Keyboard*</th><th colspan="2">Formatting PowerPoint® documents</th></tr>
<tr><td>F2 (Mac: ⌘+U)</td><td colspan="2">enter text region to edit text</td></tr>
<tr><td>TAB</td><td colspan="2">go to next object</td></tr>
<tr><td>SHFT+TAB</td><td colspan="2">go to previous object</td></tr>
<tr><td>CTRL+D</td><td colspan="2">duplicate selected object</td></tr>
<tr><td>SPACEBAR, PGDN,
<right|down arrow></td><td>advance slide</td><td rowspan="4">during slide show</td></tr>
<tr><td>BACKSPC, PGUP,
<left|up arrow></td><td>previous slide</td></tr>
<tr><td>B</td><td>black screen</td></tr>
<tr><td>W</td><td>white screen</td></tr>
</table>

* On Apple® computers, the CTRL key is the command key, ⌘.

When preparing a presentation, you should create a slide template and use it for all the slides in your presentation. The sample presentation in Section 5.9 uses my slide template: simple, contrasting headings, and lots of space for information. Figure B.23 presents my Master Layout and my style guide.

Master layout

Select the Master Layout to edit the default slides. These slides establish the defaults for fonts, text areas, and the background of your slides. Adjust these to your liking and insert any graphics that should appear on every slide, such as your logo or page numbering. Similarly, you can edit the Handout Master and the Notes Master.

Creating slides

Often, when preparing slides, you will want to repeat some information on the next slide. Instead of retyping this information onto a blank slide, duplicate the slide and then delete the information you do not require. Select the slide in the left organizer pane and press CTRL+D to duplicate the slide.

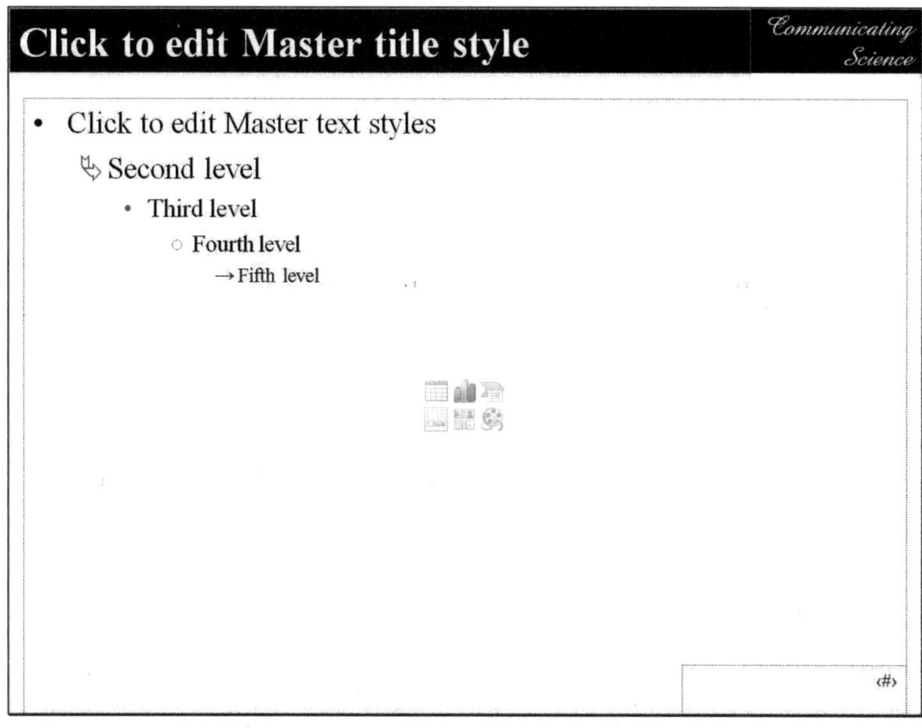

Formatting guide

Communicating Science

Heading	28 pt
General text	24 pt
General text may be indented	
Bulleted subtext	24 pt
• bulleted subsubtext	20 pt
○ ditto	18 pt
1. Numbered subtext	24 pt
1. numbered subsubtext	20 pt
comments ***comments*** *comments* ***comments***	*18 pt; blue or red; bold or not*
position of equations	indented from text
position of figures	centered
<large spacer>	24 pt
<small spacer>	10 pt

Figure B.23 The Master Layout (top) and formatting guide (bottom) used for the PowerPoint presentation in Section 5.9.

Creating a poster in PowerPoint

PowerPoint can be used to create a scholarly poster. To configure PowerPoint for your poster:

- Start with a single blank slide.
- In Page Setup on the Design ribbon, select Custom Page Size and enter the desired dimensions of your poster.*
- Adjust the font sizes on the Slide Master to the values in Section 5.7.
- Create textboxes for each text region.
  - Textboxes should have the same width, but may have varying lengths, depending on your layout.
  - To format the textbox, right-click on the textbox and then select either Size and Position... or Format Shape... to configure the textbox size, color, position, and other formatting parameters.
  - Once you have properly formatted one textbox, it may be easier to duplicate it using CTRL+D
- Ensure the textboxes are horizontally and vertically aligned, as illustrated in Figure 5.9 (page 193) and the sample posters starting on page 201. There are several ways of aligning textboxes:
  - Visualize the Gridlines and use them as an alignment guide.
  - Select the textbox edge, then adjust the location using the arrow keys to move the textbox.
- Insert figures as you would when creating an oral presentation.
  - Ensure the figures are not distorted when resized.
  - Use vector formats if possible. If the figure is raster, ensure the resized figure has greater than 200 dpi resolution, which does not show pixelation.

* The maximum poster size in PowerPoint is 142 cm × 142 cm (56" × 56"). If your poster exceeds this size, divide all the dimensions in half, including the font sizes, and then ask the printer to print the poster at 200 % magnification.

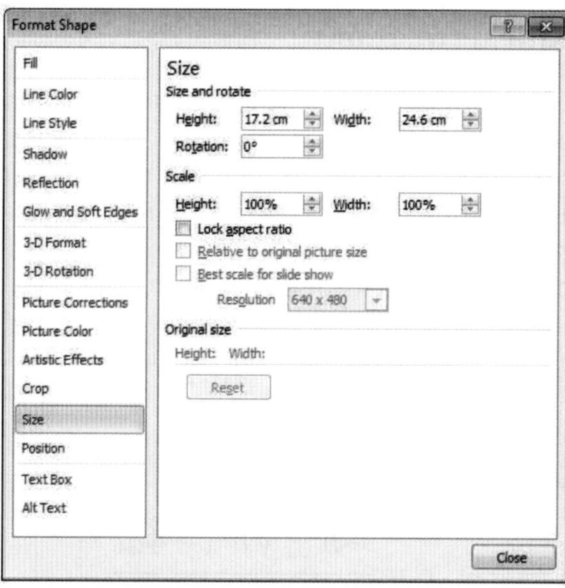

Figure B.24 The Size tab in the Format Shape textbox dialog box.

B.4 Mathematical equations

The Equation Tools add-in allows you to create equations with proper mathematical syntax in all Microsoft Office® products. To create an equation, select Equation:Insert New Equation on the Insert ribbon. This creates a blank equation and opens the Equation Tools ribbon. The default is to create the equation as a floating object. By clicking on the equation, you can change this to an inline object.

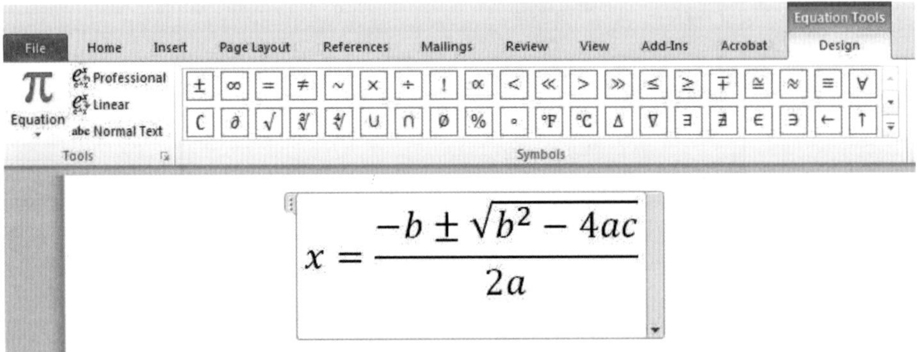

Figure B.25 The interface of the Equation Tools editor and a sample equation.

On the right-hand side are mathematical structures, from fractions to mathematical operators. On the left-hand side are common mathematical symbols.*

The Equation Tools add-in has sufficient functionality to create most common mathematical equations. Most of the equations in *Communicating Science* could be created using Equation Tools. Below are a few more examples.

$$\begin{bmatrix} a & b \\ c & d \end{bmatrix} \qquad \xrightarrow{\Delta} \qquad \int_0^\infty e^{-ax^2}\,dx = \frac{1}{2}\sqrt{\frac{\pi}{a}} \qquad {}^{20}_{10}\mathrm{Ne}$$

Figure B.26 Selected mathematical expressions that can be created using the Equation Tools.

When creating equations for large displays or PowerPoint presentations: create the equation using the normal font sizes, right-click on the equation, and then select Size and Position… to scale the image to the desired size. In PowerPoint, make the equation font the same as the normal PowerPoint font.

* If you are creating many formulae, I recommend purchasing MathType. It has more functionality and a more convenient interface.

B.5 Electronic annotation

Annotating electronic documents is increasingly common since more resources are distributed electronically. However, it is often quicker to annotate printed documents than to annotate electronic documents. Section 6.2 provides information on hand annotation.

Annotating PDF files

The software available to annotate PDF files is dependent on the operating system and changes regularly. Selected programs are listed below. They have different features and different functionality. I recommend you read the online reviews and test several programs to find one that provides both the functionality you require and that you are comfortable using.

Windows

Adobe Reader
Foxit Reader
PDF-XChange Viewer
Sumatra

Mac

Preview
Adobe Reader
Skim

iPad

Adobe Reader
iAnnotate PDF
Notability
Note Taker HD
PDF Expert

Android

Adobe Reader
ezPDF Reader
iAnnotate PDF
RepliGo

Linux

<No good native software. Use Windows software within Linux WinE.>

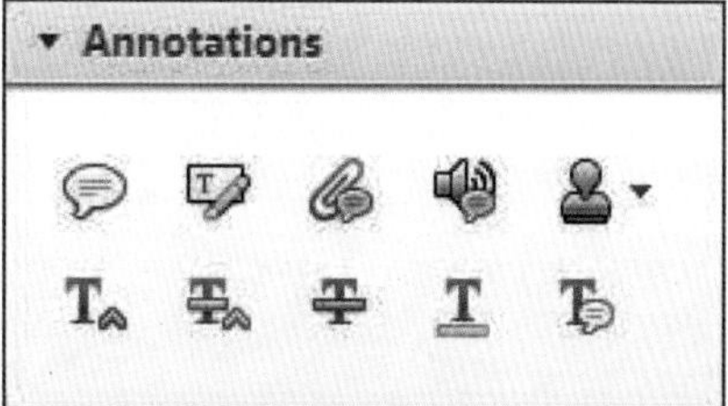

Figure B.27 The annotation tools in Adobe Reader. In order to use these tools, the author must have enabled commenting when creating the PDF.

Annotating Word® files

Before submitting your electronic document for review, select Tools:Track Changes to make Word record all changes to your document. Also select Tools:Protect Document…, then allow only Tracked Changes, and then Start Enforcing Protection. It is optional to enter a password, but a password ensures that the reviewers do not accidentally turn off Track Changes or otherwise edit the document.

The Track Changes mode causes Word to display the Reviewing toolbar in Figure B.29 and display a record of every edit made.

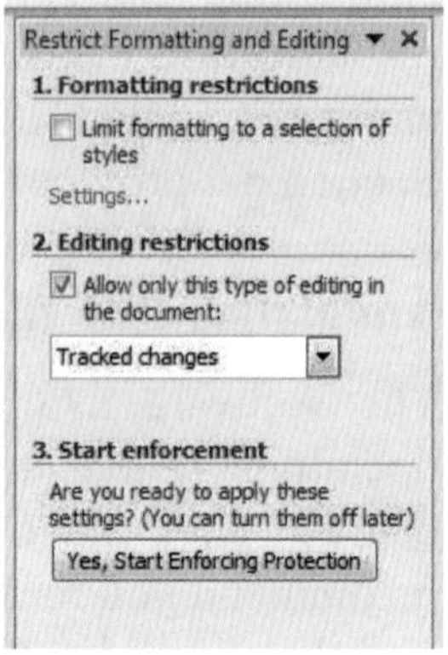

Figure B.28 The settings to protect a document prior to giving it to a reviewer.

Reviewers may do two things in the protected document:

- edit the document (add, delete, and format text)
- insert comments

Word assigns each reviewer a different color and uses different shades for edits and comments. When editing, new text appears in color and is underlined. Deleted text appears to the right of where it was deleted.[*] Selecting insert comment, ▭, will insert a comment at the location of the cursor. If text is selected, the selected text is highlighted in color and the comment box appears at right.

Figure B.29 The reviewing ribbon. Hovering over the icons explains their function.

[*] This is the Final:Showing Markup mode. In Original:Show Markup mode, deleted text is crossed out and new text appears to the right.

Accepting or rejecting changes

Once you receive the file from the reviewer, you must review every change and determine if you will accept or reject the change.

First, unprotect the document by selecting Tools:Unprotect Document....

Then, starting at the top of the document, select next change, , read the reviewer's recommendation, and then either select accept change, , reject change, , or move onto the next change if you wish to skip this recommendation for now.

Sample electronic annotations

Figure 6.2 (copied below) contains a section of hand-annotated text. This same text is annotated electronically using Adobe Reader in Figure B.30 and Microsoft Word in Figure B.31.

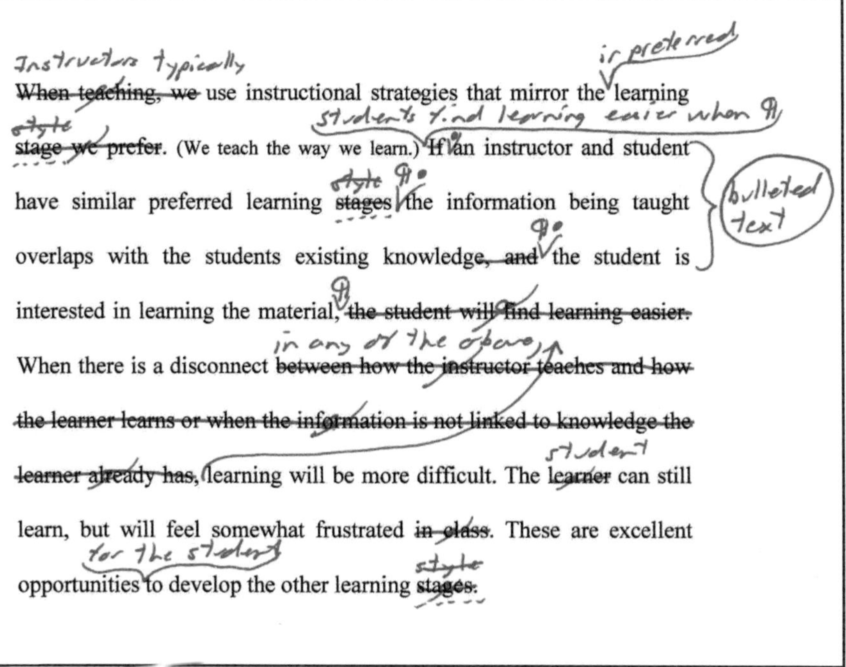

When teaching, we use instructional strategies that mirror the learning stage we prefer. (We teach the way we learn.) If an instructor and student have similar preferred learning stages the information being taught overlaps with the students existing knowledge, and the student is interested in learning the material, the student will find learning easier. When there is a disconnect between how the instructor teaches and how the learner learns or when the information is not linked to knowledge the learner already has, learning will be more difficult. The learner can still learn, but will feel somewhat frustrated in class. These are excellent opportunities to develop the other learning stages.

Figure 6.2 Reproduced for ease of comparison with Figure B.31.

After reviewing the hand-annotated version in Figure 6.2 and the electronically annotated versions in Figures B.30 and B.31, you will see that the abilities of electronic annotation tools vary. Generally, it takes longer to electronically annotate a document. Even today, many reviewers still prefer hand-annotating documents.

~~When teaching, we~~ use instructional strategies that mirror the learning stage we prefer. (We teach the way we learn.) ~~If~~ an instructor and student have similar preferred learning stages the information being taught overlaps with the students existing knowledge, and the student is interested in learning the material, ~~the student will find learning easier.~~ When there is a disconnect ~~between how the instructor teaches and how the learner learns or when the information is not linked to knowledge the learner already has,~~ learning will be more difficult. The learner can still learn, but will feel somewhat frustrated ~~in class~~. These are excellent opportunities to develop the other learning stages.

Figure B.30 Electronic annotation of the same text as in Figure 6.2 using Adobe Reader. To display new text and comments, the user must hover over each comment box and insert symbol, ▲.

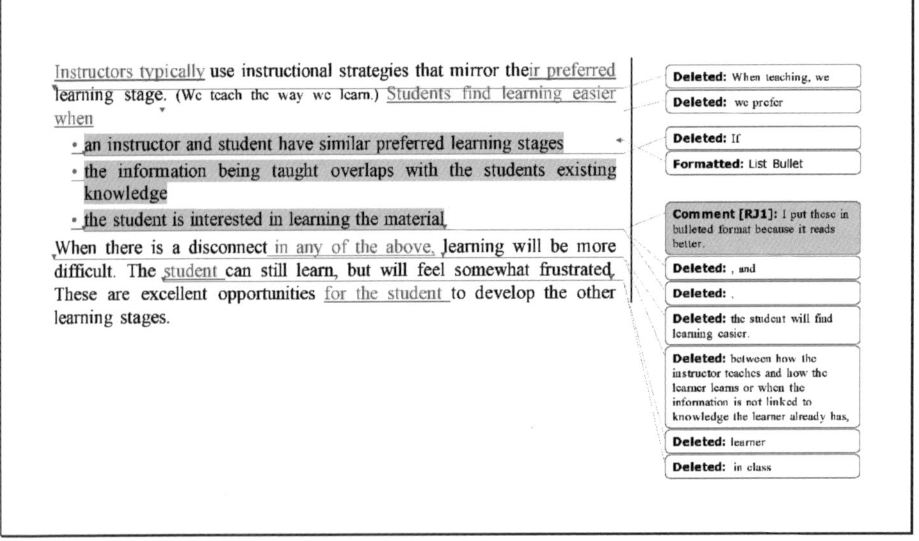

Instructors typically use instructional strategies that mirror their preferred learning stage. (We teach the way we learn.) Students find learning easier when

- an instructor and student have similar preferred learning stages
- the information being taught overlaps with the students existing knowledge
- the student is interested in learning the material

When there is a disconnect in any of the above, learning will be more difficult. The student can still learn, but will feel somewhat frustrated. These are excellent opportunities for the student to develop the other learning stages.

Deleted: When teaching, we
Deleted: we prefer
Deleted: If
Formatted: List Bullet
Comment [RJ1]: I put these in bulleted format because it reads better.
Deleted: , and
Deleted: .
Deleted: the student will find learning easier.
Deleted: between how the instructor teaches and how the learner learns or when the information is not linked to knowledge the learner already has,
Deleted: learner
Deleted: in class

Figure B.31 Electronic annotation of the same text as in Figure 6.2 using Word 2010, showing how new text, deleted text, and comments are displayed when Track Changes is on. Conveniently, Word identifies each reviewer's recommendations with colored text and the reviewer's initials in the comment boxes. It is also possible to show the original text with the changes to the right using the Original:Show Markup in Figure B.29.

Appendix C: Assignments

Practice is critical to improving your communication skills. As you practice writing, editing, and presenting, your abilities improve and your confidence increases. While the focus of *Communicating Science* is on scientific documents, the strategies developed herein can be applied to all aspects of life, from preparing your résumé to negotiating a car loan.

Below are assignments that build skills, experience, and confidence in preparing and presenting works. Most of these assignments can form the basis for in-class discussions and can be augmented with peer review and/or peer evaluation. Section 6.1 provides suggestions for establishing a safe environment where students are more willing to share openly and honestly. Instructors are encouraged to add context relevant to their instructional environment to make assignments that are meaningful to their students.

Many of these assignments ask you to use information from a science course you are taking or have taken. This strategy helps you integrate and apply your improving communication skills to a subject you are interested in. Moreover, you are learning more about the science subject!

Introductory assignments

1. Prepare a 20 – 30 second story about yourself that you will present during class. (The instructor presents first.)
2. In pairs, interview each other and prepare a 20 – 30 second story about the other person that you will present during class. (These should be template-based.)
3. Without using *Communicating Science* or other resources, prepare a list of the types of written and oral communication done by scientists. What is the purpose of each communication? Who is the intended audience?

In-class assignments

1. Given a few paragraphs from a document,
   a) examine each sentence to identify the type of sentence, clauses, and the elements in the sentence.
   b) examine each sentence to identify the voice, tense, and tone.

   Use these results for small group and classroom discussions on how to improve the prose.
2. Identify the nominalizations in Table 1.3.

3. Given a few of the undefined commonly confused word sets on page 49, define them. Do this in small groups, then present yours in class.
4. Search the internet to find examples of a cliché, idiom, tautology, circumlocution, pleonasm, and/or vague & imprecise word. Rewrite the sentence to improve readability. Share with the class.
5. Given _#_ clichés, idioms, tautology, circumlocution, pleonasms, and/or vague & imprecise words,
   a) use Google Ngram to visualize the commonness of the phrase over time.
   b) search a scholarly database to identify phrases still in use and the discipline(s) they are used in.

   Discuss why the commonly used phrases are still in use.
6. Given the individual sentences from a paragraph out of order, order the sentences to produce a coherent paragraph. Identify the topic sentence of the paragraph.
7. Given a document, identify grammatical and style errors in the document. Edit the document and recommend corrections.
8. Given a document, what revisions could be made to make the document more clear, coherent, concise, and precise?
9. Given a scholarly article, identify the key points of the article.
10. Given a scholarly article with references removed, identify sections of text that require citation.
11. Given a scholarly article with the abstract removed, prepare an abstract. Review and improve upon the abstract in the next class.
12. Given a scholarly article, write the "teaser" for a website or the front pages of the journal.
13. Given a public press article that misrepresents a scientific concept, write a letter that identifies the misrepresentation and correctly explains the science, at the level of the reader. Review and improve upon the response in the next class.
14. Given a public press article that denigrates science to further a political or corporate agenda, prepare a reasoned and learned rebuttal to submit to the editor. As a class, review, revise, and submit a response.
15. Given a science exam question, prepare a written or verbal explanation of the answer to a fellow student taking the course and to a student who has not taken the course.

16. Given a concept you are learning in another science course, write two paragraphs: one conveying that concept to a fellow student, and one conveying that concept to a grade _#_ student.
17. Find a report or essay that you wrote at the last minute for another course. (That is, you started and finished it the night before it was due.) Review this report or essay.
18. Write an instruction guide for a common activity: how to tie your shoes, how to bake cookies, etc.
19. Given a job posting, prepare a cover letter to apply for that position.
20. Given a scientific discipline, determine the resources available on the internet, from your institutional library, and/or from your public library to a) search primary literature, and b) access entire articles.

In-depth assignments

1. Create a résumé or curriculum vitae.
2. Prepare an email/letter to a professor asking about an advertised summer research opportunity.
3. Given a scholarly article, prepare a summary of the article.
4. Given a scholarly article, rewrite for a public audience with minimal loss of scientific accuracy.
5. Prepare an investigative essay on ethics, professionalism, and plagiarism. (This can serve as a basis for discussions of these topics.)
6. Prepare a research proposal. Present this to a panel of fellow students. (Option: Dragon's Den™ or Shark Tank™ concept)
7. Identify and investigate instances where private, corporate, or government results were skewed because of pressure to obtain the "preferred" results.
8. Select a scientist or graduate student researcher, interview them, and write a newspaper article about their career and achievements.
9. Select a concept that you are currently learning about in another science course.
    a) Prepare a ten-minute class presentation on the topic. (classroom concept)
    b) Prepare a ten-minute pre-laboratory presentation on the topic, possibly with a demonstration. (laboratory concept)

10. Identify a faculty member preparing a document for publication.
    a) With permission of the author(s), review the document.
    b) Have the author(s) attend an open-feedback session. Have the course instructor facilitate how to properly give and receive feedback.

Term projects

For term projects, it is valuable to establish submission dates for project milestones, such as topic selection, project outline, annotated bibliography, and peer review events. The document should be complete, not draft, for all of the peer review events.

1. Adapt one of your laboratory reports into a scholarly poster, article, and/or presentation.* (Use different laboratory reports for each.)
2. Prepare an investigative essay on
    a) a sustainability or environmental issue from the perspective of your discipline.
    b) a controversial scientific claim or possible pseudoscience claim.
    c) the science underlying a real-world application of science.
    d) a concept you are learning in another science course.
3. Prepare a scholarly poster for your essay from question 2, above.
4. With the permission of the laboratory coordinator, assign sections of the laboratory manual to groups of students. The groups edit the laboratory manual, consulting the laboratory coordinator as required. The instructor facilitates reviews by other student groups. With appropriate permissions, the revised experiments could be used in future editions of the laboratory manual!†
6. In consultation with the course instructor, select a scholarly publication at the beginning of the term. This article forms the basis for several assignments: a technical article (science news article), a non-technical article (newspaper article), a review article of progress since this particular article, a poster presentation, and an oral presentation, etc.

* This assignment provides an opportunity for you to learn more about a subject you are interested in, learn how to better communicate that material, and possibly provide the department with information to improve the experiment.

† This assignment is not a make-work project. It has real outcomes. The institution obtains an improved document. The students are published and have something to add to their résumé. A win-win assignment.

Appendix D: Assessment rubrics

Grading rubrics for laboratory reports, scholarly articles and essays, scholarly posters, and scientific presentations are given below.* There are three categories of criteria: content, organization, and presentation. These rubrics have a maximum of 30 points. Also included are a rubric for reviewing the members on your team (maximum 20 points) and for reviewing the reviewer (maximum 9 points).

About scholarly articles: the variability between disciplines makes it challenging to put together a rubric for grading a scholarly article. One option would be to enlist academic colleagues in the appropriate disciplines to evaluate student articles.

Engaging students

It may be challenging to engage students in meaningful peer-review and peer-evaluation — meaningful in that students put in the effort to accurately assess the work of their peers. The information on page 248 (Peer review in a classroom) sets the context for quality and effective peer review. Some students may want to give all their peers full marks. To address this, emphasize the following:

> If every student gets an A, the grade is meaningless. Consider exams: while the grades range from 0 to 100 %, the average is typically between 60 and 80 %. For this project: you can assess any individual project from 0 to 100 % using the rubric, but your average must be between 60 and 80 %.

Some students may want to give all their peers the same grade. This penalizes those who did well and inflates the grades of those who put little effort into their project. Informing students why this is detrimental to learning is often enough to convince most students to accurately assess their peers. Alternatively, to emphasize the importance of meaningful peer review and to encourage students to develop these skills, it is possible to impose a grade penalty:

> You are expected to critically and impartially assess your peers. 20 % will be deducted from your grade if your peer-evaluation average is not in the 60 – 80 % range or if all your grades are within ± 20 % of each other.

* An internet search for "science rubric" returns a plethora of rubrics. Many of them are duplicates of each other and there is no indication of the original author. The rubrics in this section were adapted from those found online, but it is not practically possible to identify and credit the original authors. (See page 89.)

LAB. REPORT	Excellent ③	Good ②	Poor ①	Unsatisfactory ⓪
content				
Introduction & theory	Presents a clear and concise introduction to the topic.	Presents too much information on the topic.	Presents insufficient information.	Is too short or superficial.
Procedure	Easy-to-follow steps which are logical and sufficiently detailed.	Most steps are easy-to-follow. Some are unclear or lack detail.	Some of the steps cannot be understood.	Procedure not consistent with experiment.
Data and observations	Data presented clearly and concisely. Significant figures correct. Important observations recorded.	Data presentation could be improved. Some observations missing.	Extraneous data included, or no observations.	Incomplete or inaccurate.
Calculations	All calculations are correct, clearly presented, and include dimensional analysis.	Most calculations are correct and clearly presented, and include dimensional analysis.	Most calculations are correct and clearly presented. Dimensional analysis incorrect or absent.	Many calculations are incorrect or absent.
Discussion	Theory properly applied to interpret data. Logical and thorough discussion shows understanding.	Theory properly applied to interpret data. Discussion similar to introduction.	Inconsistent application of theory to interpretation. Discussion suggests limited understanding.	Unclear and/or illogical.
Conclusions	Logical conclusions are drawn from the data. Impact to science and society are reasonable.	Logical conclusions are drawn from the data. Impact to science and society are limited.	More conclusions could be drawn from the data.	Conclusions are not consistant with the data.
Language	Prose is clear, concise, and at the appropriate readability level for the audience.	Minor lapses in clarity, concision, or readability.	Major lapses in clarity, concision, or readability.	One or more sections of the document are not understandable by the intended audience.
Figures and tables	Figures and tables are focused, on-topic, and easy to read.	Figures and/or tables contain extraneous information.	Figures and/or tables are difficult to read.	Figures and/or tables are missing or contain inaccurate information.
presentation				
Style	Follows the style guide. Aesthetically appealing.	A few deviations from the style guide.	Several deviations from the style guide. Inconsistent formatting.	Does not follow the style guide. Poor aesthetic presentation.
Spelling and grammar	No spelling or grammatical errors.	A few minor spelling or grammatical errors.	Spelling or grammatical errors affect the readability.	Spelling or grammatical errors make sections unintelligible.

ARTICLE / ESSAY	Excellent ③	Good ②	Poor ①	Unsatisfactory ⓪
content				
Introduction	Presents a clear and concise introduction to the topic.	Presents too much information on the topic.	Presents insufficient information.	Is too short or superficial.
Body of work	Information is focused, in-depth, and on-topic.	Information is in-depth, but strays into related topics.	Information covered superficially.	Topic is too broad and/or covered superficially.
Accuracy	All information is accurate and in-depth. No factual errors.	All information is accurate, but ocassionally superficial.	Some inaccurate and/or superficial information.	Significant inaccurate information.
Conclusions	Logical conclusions are drawn from the work. Impact to science and society are reasonable.	Logical conclusions are drawn from the work. Impact to science and society are limited.	More conclusions could be drawn from the work.	Conclusions are not consistant with the work.
Language	Prose is clear, concise, and at the appropriate readability level for the audience.	Minor lapses in clarity, concision, or readability.	Major lapses in clarity, concision, or readability.	Sections of the document are not understandable by the intended audience.
Figures and tables	Figures and tables are focused, on-topic, and easy to read. Calculations are correct.	Figures and/or tables contain extraneous information.	Figures and/or tables are difficult to read.	Figures and/or tables are missing or contain inaccurate information.
organization				
Continuity	Logical organization of detailed information presented in a cohesive manner.	A different order would better present the information.	Some discontinuity between topics.	Poor continuity between topics.
Citations	Information is from reputable sources and cited properly. Possible bias is documented.	Information is mostly from reputable sources.	Extra citations are included.	Sources are questionable and/or citations are missing.
presentation				
Style	Follows the style guide. Aesthetically appealing.	A few deviations from the style guide.	Several deviations from the style guide. Inconsistent formatting.	Does not follow the style guide. Poor aesthetic presentation.
Spelling and grammar	No spelling or grammatical errors.	A few minor spelling or grammatical errors.	Spelling or grammatical errors affect the readability.	Spelling or grammatical errors make one or more sections unintelligible.

POSTER	Excellent ③	Good ②	Poor ①	Unsatisfactory ⓪
content				
Information	Introduction is concise and engaging. Information is focused, in-depth, and on-topic.	Introduction is concise and engaging. Information is in-depth, but strays into related topics.	Introduction fails to rationalize project. Information covered superficially.	Topic is too broad and/or covered superficially.
Accuracy	All information is accurate and in-depth. No factual errors.	All information is accurate, but ocassionally superficial.	Some inaccurate and/or superficial information.	Significant inaccurate information.
Conclusions	Logical conclusions are drawn from the data. Impact to science and society are reasonable.	Logical conclusions are drawn from the data. Impact to science and society are limited.	More conclusions could be drawn from the data.	Conclusions are not consistant with the data.
Language	Prose is clear, concise, and at the appropriate readability level for the audience.	Minor lapses in clarity, concision, or readability.	Major lapses in clarity, concision, or readability.	Sections of the document are not understandable by the intended audience.
Figures and tables	Figures and tables are focused, on-topic, and easy to read. Calculations are correct.	Figures and/or tables contain extraneous information.	Figures and/or tables are difficult to read.	Figures and/or tables are missing or contain inaccurate information.
organization				
Continuity	Logical organization of detailed information presented in a cohesive manner.	A different order would better present the information.	Some discontinuity between topics.	Poor continuity between topics.
Citations	Information is from reputable sources and cited properly. Possible bias is documented.	Information is mostly from reputable sources.	Extra citations are included.	Sources are questionable and/or citations are missing.
presentation				
Appearance	Catchy, attracts visitors. All text readable from 1 m; title from 3 m.	Pleasant appearance. Text in figures and/or tables difficult to read.	Looks messy. Some sections difficult to read.	Looks busy. Difficult to follow and read.
Spelling and grammar	No spelling or grammatical errors.	A few minor spelling or grammatical errors.	Spelling or grammatical errors affect the readability.	Spelling or grammatical errors make sections unintelligible.
Presentation and questions	Presenter confident, interesting, and engaging. Presenter understands work in detail.	Presenter lacks confidence, but understands work.	Presenter has limited understanding of work.	Presentation memorized or read. Poor pronounciation, poor projection, minimal eye contact.

PRESENTATION	Excellent ③	Good ②	Poor ①	Unsatisfactory ⓪
content				
Information	Introduction is concise and engaging. Information is focused, in-depth, and on-topic.	Introduction is concise and engaging. Information is in-depth, but strays into related topics.	Introduction fails to rationalize work. Information covered superficially.	Topic is too broad and/or covered superficially.
Accuracy	All information is accurate and in-depth. No factual errors.	All information is accurate, but ocassionally superficial.	Some inaccurate and/or superficial information.	Significant inaccurate information.
Conclusions	Logical conclusions are drawn from the data. Impact to science and society are reasonable.	Logical conclusions are drawn from the data. Impact to science and society are limited.	More conclusions could be drawn from the data.	Conclusions are not consistant with the data.
Figures and tables	Figures and tables are focused, on-topic, and easy to read. Calculations are correct.	Figures and/or tables contain extraneous information.	Figures and/or tables are difficult to read.	Figures and/or tables are missing or contain inaccurate information.
organization				
Continuity	Logical organization of detailed information presented in a cohesive manner.	A different order would better present the information.	Some discontinuity between topics.	Poor continuity between topics.
Slide layout	Headings, text, and visuals are easy to read and logically laid out on slides. One topic per slide.	Structured layout, but appears cluttered or busy.	Hard to read because of font, color, or layout.	Few headings, poor organization. Too much information per slide.
Citations	Information is from reputable sources and cited properly. Possible bias is documented.	Information is mostly from reputable sources.	Extra citations are included.	Sources are questionable and/or citations are missing.
presentation				
Multimedia	Multimedia is focused on-topic, concise, and aids understanding.	Multimedia is focused on-topic.	Some multimedia is not relevant to the topic *or* animations detract from presentation.	Multimedia is lengthy and/or detracts from topic.
Presentation	Presenter confident, interesting, and engaging. Projects voice and makes regular eye contact.	Presenter lacks confidence, but makes attempts to project voice and make eye contact.	Presenter adds little additional information than what is on slide.	Presentation memorized or read. Poor pronounciation, poor projection, minimal eye contact.
Questions	Answers questions confidently and accurately. Elaborates on material in presentation.	Answers questions by repeating presentation material.	Hesitantly answers questions. Uncertain about material.	Superfical answers, or doesn't answer questions.

TEAMWORK	Excellent ④	③	Good ②	①	Unsatisfactory ⓪
Contribution	Does more or higher-quality work than expected. Keeps the team organized. Has their assignments completed early. Helps teammates who are having difficulty completing their work.	Demonstrates behaviors described in both ② and ④.	Completes a fair share of the team's work with acceptable quality. Keeps commitments and completes assignments on time. Helps teammates who are having difficulty when it is easy or important.	Demonstrates behaviors described in both ⓪ and ②.	Does not do a fair share of the team's work. Delivers sloppy or incomplete work. Misses deadlines. Is late, unprepared, or absent from team meetings. Does not assist teammates. Quits if the work becomes difficult.
Expectations	Motivates the team to do excellent work. Is enthusiastic about the project. Wants the team to do outstanding work, even if there is no additional reward. Believes that the team can do excellent work.		Encourages the team to do good work that meets all requirements. Wants the team to perform well enough to earn all available rewards. Believes that the team can meet all of its responsibilities.		Satisfied, even if the team does not meet assigned standards. Wants the team to avoid work, even if it hurts the team. Doubts that the team can meet its requirements.
Knowledge and skills	Demonstrates the knowledge, skills, and abilities to do excellent work. Acquires new knowledge or skills to improve team performance. Performs the role of any team member if necessary.		Demonstrates sufficient knowledge, skills, and abilities to contribute to the team's work. Acquires knowledge or skills as needed to meet requirements. Can perform some of the tasks normally done by other teammates.		Missing basic qualifications needed to be a member of the team. Unable or unwilling to develop knowledge and skills to contribute to the team. Unable to perform any of the duties of other team members.
Relationships	Interested in teammates' ideas and contributions. Makes sure teammates stay informed and understand each other. Asks teammates for feedback and uses their suggestions to improve.		Listens to teammates and respects their contributions. Communicates clearly. Shares information with teammates. Respects and responds to feedback from teammates.		Interrupts, ignores, or ridicules teammates. Takes actions that affect teammates with-out their input. Does not share information. Is defensive. Will not accept help or advice from teammates.
Team focus	Monitors team dynamics and team progress. Ensures the team is progressing at a reasonable pace. Gives teammates specific, timely, and constructive feedback.		Notices changes in dynamics that affect team progress. Knows what teammates should be doing and identifies problems. Alerts the team to possible problems that may affect the team success.		Is unaware of team progress. Does not pay attention to the progress of their teammates. Avoids discussing obvious team problems.

adapted from a rubric by Dr. Matthew Ohland, et al., Engineering, Purdue University

Reviewer	Excellent ③	Good ②	Poor ①	Unsatisfactory ⓪
Comprehensive	The reviewer provided extensive feedback on the entire document.	The reviewer provided minimal feedback on some sections of the document.	The reviewer provided no feedback on some sections of the document.	The reviewer provided no feedback on major sections (more than half) of the document.
Quality	The reviewer identified all areas that required major and minor improvement.	The reviewer identified most areas that required major and minor improvement.	The reviewer identified most areas that required major improvement.	The reviewer missed sections that obviously required improvement.
Feedback	The reviewer provided constructive feedback.	The reviewer's feedback was mostly constructive with a few negative statements.	The reviewer's feedback was a mix of constructive and negative statements.	The reviewer's comments were negative or the reviewer provided little useful feedback.

PRESENTATION	strongly disagree 1	disagree 2	neutral 3	agree 4	strongly agree 5
The presentation is focused and on-topic.					
The presentation is at an appropriate scientific level.					
The material is presented in a way that facilitates learning. (logical progression, appropriate instructional strategies)					
The slides contain a mix of text and images that stay on topic. (appropriately sized text and images, headings and captions appropriate)					
The presentation cites appropriate scientific sources.					
The speakers are audible and convey enthusiasm about the topic.					
The speakers are knowledgeable about the material. (didn't read script, correct pronounciation, have greater knowledge than is presented)					
The presentation is reasonably paced and flows smoothly. (practiced and of professional quality)					
Questions are answered professionally.					
Overall, the presentation is interesting and informative.					

Points: __________ **/50**

POSTER	strongly disagree 1	disagree 2	neutral 3	agree 4	strongly agree 5
The poster is focused and on-topic.					
The poster content is at an appropriate scientific level.					
The poster is accurate and presented in a way that facilitates learning. (short statements, bulleted points, graphics)					
The poster layout is logical and easy to follow. (short introduction, results and discussion, short summary/conclusion)					
The poster contains a mix of text and images that stay on topic. (appropriately sized text and images, headings and captions appropriate)					
The poster cites appropriate scientific sources.					
The speakers are audible and convey enthusiasm about the topic.					
The speakers are knowledgeable about the material. (didn't read script, correct pronounciation, have more knowledge than on poster)					
The presentation is reasonably paced and flows smoothly. (practiced and of professional quality)					
Questions are answered professionally.					

Points: __________ **/50**

Index

— H —

— I —

— J —

— K —

— L —

— M —

— N —

— O —

— P —